SCALING OF WATER FLUX RATE IN ANIMALS

# Scaling of Water Flux Rate in Animals

by Kenneth A. Nagy and
Charles C. Peterson

UNIVERSITY OF CALIFORNIA PRESS
Berkeley • Los Angeles • London

UNIVERSITY OF CALIFORNIA PUBLICATIONS IN ZOOLOGY

Editorial Board: Peter B. Moyle, James L. Patton,
Donald C. Potts, David S. Woodruff

Volume 120
Issue Date: July 1988

UNIVERSITY OF CALIFORNIA PRESS
BERKELEY AND LOS ANGELES, CALIFORNIA

UNIVERSITY OF CALIFORNIA PRESS, LTD.
LONDON, ENGLAND

ISBN 0-520-09738-6
LIBRARY OF CONGRESS CATALOG CARD NUMBER: 88-10791

© 1988 BY THE REGENTS OF THE UNIVERSITY OF CALIFORNIA
PRINTED IN THE UNITED STATES OF AMERICA

Library of Congress Cataloging-in-Publication Data
Nagy, Kenneth A.
    Scaling of water flux rate in animals.

    (University of California publications in zoology;
v. 120)
    "A preliminary report of the results of this analysis
was presented at the Eighth International Conference on
Comparative Physiology, Carns-Sur-Sierre, Switzerland,
in June 1986"—Acknowledgments.
    Bibliography: p.
    1. Water in the body. 2. Osmoregulation.
3. Physiology, Comparative. I. Peterson, Charles C.
II. Title. III. Series.
QP90.5.N34    1988      599'.019212      88-10791
ISBN 0-520-09738-6 (alk. paper)

# Contents

*List of Figures,* vii
*List of Tables,* ix
*Acknowledgments,* x
*Abstract,* xi

INTRODUCTION ..... 1

METHODS ..... 3

SCALING IN MAJOR TAXONOMIC GROUPS ..... 7
   Eutherian Mammals, 7
   Marsupial Mammals, 10
   Birds, 10
   Reptiles, 11
   Amphibians, 13
   Fishes, 13
   Arthropods, 14
      Water-breathing, 14
      Air-breathing, 15
   Mollusks, 16
   Annelids, 16
   Single Cells, 17
   Summary, 17

COMPARATIVE SCALING ..... 18
   In Captivity, 18
   In the Field, 18

DIET AND HABITAT EFFECTS 20
    Eutherians in the Field, 21
    Marsupials in the Field, 23
    Birds in the Field, 24
    Reptiles in the Field, 24
    Amphibians in Captivity, 25
    Fishes in Captivity, 26
    Arthropods in Captivity, 29

PREDICTING WATER FLUX RATE 30

WATER ECONOMY INDEX 32
    Endotherms and Ectotherms, 33
    Desert and Nondesert Animals, 35
    Body Mass Effect, 35
    WEI Interpretation, 36

CONCLUSIONS 38

*Appendices,* 41
*Literature Cited,* 131

# List of Figures

1. Allometry of water flux rate among eutherian mammals — 8
2. Allometry of water flux rate among marsupial mammals — 9
3. Allometry of water flux rate among birds — 11
4. Allometry of water flux rate among reptiles — 12
5. Allometry of water flux rate among fishes — 13
6. Allometry of water flux rate among water-breathing arthropods — 14
7. Allometry of water flux rate among air-breathing arthropods — 15
8. Allometry of water flux rate among captive mollusks — 16
9. Comparison of allometric regressions for water flux among major taxonomic groups; wild animals in captivity and domesticated animals — 17
10. Comparison of allometric regressions for water flux among terrestrial vertebrate taxa; wild animals living in their natural habitats — 19
11. Allometry of water flux among free-living eutherian mammals according to differences in (a) diet and (b) habitat — 21
12. Allometry of water flux among herbivorous and carnivorous marsupials in the field — 22
13. Allometry of water flux in free-living birds in relation to (a) taxon, (b) diet, and (c) and (d) habitat — 23
14. Allometry of water flux in free-living desert and nondesert reptiles — 25
15. Allometry of water flux among osteichthian and chondrichthian fishes in sea water and fresh water in captivity — 26

16. Influence of water salinity on allometry of water flux in water-breathing arthropods in the laboratory     27
17. Allometry of water flux in captive, air-breathing arthropods in relation to habitat     28
18. Water economy index (the amount of water used per kilojoule of energy metabolized by animals) for air-breathing animals maintaining constant body masses in the field     34
19. Relationship of water economy index (WEI) in free-living, steady-state animals to body mass     36

# List of Tables

1. Summary of equations for water flux rate in animals — 31
A1. Summary of water flux rates, measured using tritiated or deuterated water, in captivity (including domestic animals) and in the field — 42
A2. Regression statistics for allometry of water flux rate in animals — 120
A3. Summary of water economy index (WEI) values (ratios of water flux to metabolic rate) in free-living animals — 124

# Acknowledgments

This review, and many of the unpublished results reported herein, received funding in whole or part from U.S. Department of Energy Contract DE-AC03-76-SF00012 and from National Science Foundation Grant DEB 79-03868. We are grateful to our colleagues A. J. Bradley, S. D. Bradshaw, B. Clay, A. A. Degen, H. I. Ellis, M. R. Fleming, W. J. Foley, P. Greenaway, A. Hazan, N. K. Jacobsen, M. Kam, W. W. King, A. K. Lee, R. W. Martin, K. D. Morris, B. S. Obst, T. N. Pettit, D. R. Powers, I. J. Rooke, G. Sanson, G. C. Whittow, and R. P. Wilson for permission to cite as-yet-unpublished results, to two anonymous referees for their excellent suggestions for revision, and to M. Gruchacz, B. Henen, and A. Roberts for their help in various capacities with this project. A preliminary report of the results of this analysis was presented at the Eighth International Conference on Comparative Physiology, Crans-Sur-Sierre, Switzerland, in June 1986.

# Abstract

Rates of water flux, all measured using tritiated or deuterated water in captive and free-living eutherian mammals, marsupial mammals, birds, reptiles, fishes, water- and air-breathing arthropods, and mollusks, were summarized and analyzed allometrically ($\log_{10}$-$\log_{10}$ regressions). Water flux is strongly correlated with body mass within major taxa, and it scales differently (ANCOVA) in captive vs. free-living animals, among animals in different taxa, and among animals having different habitats and diets. This indicates that ecological conclusions based on laboratory studies of captive animals are suspect. Water fluxes are highest in water-breathing animals, intermediate in air-breathing endotherms (mammals and birds), and lowest in air-breathing ectotherms (reptiles and arthropods). Analyses of the "water economy index" (WEI, the amount of water used per unit of energy metabolized, as determined using doubly labeled water in free-living animals) indicate that the low water fluxes of air-breathing ectotherms (compared to endotherms) are primarily due to their low metabolic rates, rather than to supposedly more effective water economies. Desert ectotherms and endotherms have lower water fluxes and WEI values than their nondesert relatives, reflecting the importance of physiological, behavioral and morphological adaptations for achieving water balance in desert animals living under natural conditions.

Key words: allometry; captivity effect; desert; deuterated water; doubly labeled water; osmoregulation; tritiated water; water balance; water economy index.

Topic sentences: water flux scaling in animals; body size effects on water use; models of animal water flux; habitat, diet and taxon effects on water metabolism.

# INTRODUCTION

Animals are made up primarily of water. They are not closed systems, but are continuously exchanging water with their environment. Yet they must maintain a relatively constant amount of water in their bodies in the face of a variety of environmental challenges that may include water excess or its extreme unavailability. The colonization of land was accompanied by major changes in the availability of water to animals. As progressively drier habitats were invaded, animals had to find ways either to obtain increasingly scarce water from their environment, to reduce water needs by restricting water losses, to tolerate temporary imbalances until water became available, or to achieve some functional combination of these feats.

The extent to which various animals have achieved a reduced dependence on the availability of environmental water is reflected by the amount of water they process through their bodies each day: relatively independent animals have low water flux rates. The daily rate of water flux can readily be measured by labeling an animal's body water with trace amounts of the hydrogen isotopes deuterium or tritium, and then following the washout kinetics of the isotope (see Methods, below).

Water flux rates in animals that breathe water should be much higher than those in animals that breathe air. Respiratory surfaces must be permeable to respiratory gases, and thus to water, which is a comparatively small molecule. Diffusional exchange of water across respiratory membranes is measured by the labeled water method, and this avenue of water exchange overwhelms water taken in via the avenues of eating and drinking in water-breathing animals. Thus, we expect the intercepts (elevations) of the scaling relationships for water-breathing animals to be much higher than those for air-breathing animals.

Water exchange across the surfaces of air-breathing animals should be low because, compared to an equal volume of water, air contains very little water. In fact, water vapor exchange accounts for only a small fraction of total water input in typical terrestrial vertebrates (Nagy and Costa 1980), and vapor input may be negligible in deserts or anywhere else the air is dry. For terrestrial animals, the water in their food (dietary succulence), and either drinking or absorption of liquid water, become the major avenues of water intake. The availability of drinking water differs in different habitats, and the amount of water contained in different food types (e.g., animal matter, green vegetation) also varies (Shoemaker and Nagy 1977), so water flux rates in free-living, air-breathing animals might be expected to sort out on the bases of habitat and diet.

Water flux rates have now been measured in hundreds of species of animals, under a variety of conditions, in captivity and in the field. We have gathered all data available to date (May 1986) on isotopically measured water fluxes, and have used allometric analysis and analysis of covariance to summarize these data and to evaluate several hypotheses. Many physiological rate processes in animals are exponentially related to body mass ("allometry," or "scaling"), and allometric analyses have been used to test a variety of hypotheses and answer many kinds of questions about animals (see reviews by Peters 1983, Calder 1984, Schmidt-Nielsen 1984). Our questions include: (1) Does water flux rate scale with body mass within major taxonomic groups? (2) If so, does water flux scale differently in captive animals as compared with free-living animals in the same taxon? (3) Does field water flux decline with increasing aridity of habitat? (4) Does field water flux scale differently in animals having different diets or microhabitats within taxa?

Finally, we analyze field water flux data in relation to the metabolic rates of animals by calculating water use as a function of energy expenditure (the ratio of ml water used per kJ of energy used, termed "water economy index," or WEI). This ratio should be independent of general body size and taxonomic effects on rate processes, thereby making possible direct comparisons of the water economy of, for example, a small lizard, a large mammal, and a scorpion. The allometric relationships are most valuable for use in predicting water flux rates from knowledge of body mass, the WEI values best reflect water conservation adaptations, and both together are needed to evaluate the means animals use to maintain water balance in their natural habitats.

# METHODS

This review includes only water flux measurements made with deuterated or tritiated water. In an animal with labeled body water, the hydrogen isotope concentration declines exponentially through time because of the simultaneous loss of labeled water from the animal and the gain of unlabeled water (1) from the environment and (2) from production of new water via metabolic processes. Isotopic water measurements of water flux are accurate to within about ± 8% in captive animals (Nagy and Costa 1980), and we expect errors in field studies of free-ranging animals to be within ±10%.

"Water flux" is the term we used to signify the amount of water moving into an animal each day, in units of ml/d. When an animal is maintaining water balance, body water volume remains constant and the rate of water influx equals the rate of water efflux. In this situation, the washout rate of the hydrogen isotope is proportional to the flux rate of water through the animal ("water turnover" in older publications). However, when the animal's body water volume is increasing or decreasing, water influx cannot equal water efflux, and the terms "water flux" and "water turnover," which imply a steady-state condition, are misleading. Moreover, when the animal is not in a steady-state, the washout rate of the hydrogen isotope is not directly proportional to either water influx rate or water efflux rate, but lies somewhere in between. Measurements of body water volume at the beginning and end of the measurement period are required in order to calculate correct water influx and efflux rates in nonsteady-state situations (Nagy and Costa 1980). Many authors have not recognized this potentially large error, and their articles include no mention of whether their animals had constant or changing water volumes or body masses.

Fortunately, water flux rates calculated with the steady-state equation are relatively close to actual water influx rates (but are far from actual water efflux rates) in most nonsteady-state situations (Nagy and Costa 1980). Thus, by expanding our definition of water flux in this study to include water influx in nonsteady-state situations, we were able to include many more articles in this review. Some studies of arthropods, fish and aquatic reptiles, where only water efflux values were reported, are also included. These errors have little effect on our scaling calculations because they should distribute randomly about the regression lines and have minimal effect on the slope and intercept, and because the typical 10-20% errors become much smaller in both a relative and a statistical sense after logarithmic transformation.

A much more important error in some articles is the incorrect assumption that biological half-life ($T_{1/2}$) of the isotope is the same as the amount of time required for an animal to turn over half of its body water volume. The relationship between isotope kinetics and isotope turnover rate ($k$) is given by the equation $k = \ln(H_1^*/H_2^*)/t$, where $k$ is fractional turnover rate of the isotope (and body water under steady-state conditions), $\ln(H_1^*/H_2^*)$ is the natural logarithm of the ratio of initial hydrogen isotope concentration to final isotope concentration, and $t$ is time (usually in days) elapsed between initial and final measurements. The half-life of the hydrogen isotope is related to $k$ according to the equation $T_{1/2} = 0.693/k$. When $H_2^*$ is half of $H_1^*$, half of the isotope has left the animal and one half-life has elapsed. The numerator in the first equation becomes ln 2, which equals 0.693, and $t = T_{1/2}$. As an example of the error involved in equating $T_{1/2}$ with body water half-life, assume that an animal washes out half of its H* in one day. The incorrect interpretation is that daily water flux was 50% of body water volume. In fact, actual water flux was 69.3% of body water volume. When we detected this error in our literature search, we corrected the water flux values appropriately. Several authors reported $k$ and/or $T_{1/2}$, but did not report total body water (TBW) volume, which is needed for calculation of water flux in ml/d. In these cases, we used mean values of TBW for the species from other studies. For those articles wherein body mass was not given, we again used mean values for the species from other studies, and indicated this in Table A1 with a superscript.

We sorted the available water flux data, summarized in Table A1 (see Appendix), into the following taxonomic groups: eutherian mammals, marsupial mammals, birds, reptiles, amphibians, fishes, arthropods (separated into those that breathe air and those that breathe water), mollusks, polychaete worms, and single cells. Within these groups, the data were further separated into two categories: (1) wild animals studied while in captivity, and domestic animals; and (2) wild animals studied while living in their natural habitats.

We attempted to include in this review all available studies incorporating isotopically measured water fluxes in animals. However, in some studies where water fluxes were reported for a large number of experimental protocols, we selected only some of the data for analysis herein. We chose data from experimental conditions that appeared to be nonstressful (e.g., control values) as well as the highest and lowest water fluxes reported, unless these values were from animals that apparently were close to dying. Our intent was to obtain data sets for captive animals that reflect the variation in water flux rates that animals can achieve. Assessing the range of physiological capabilities possessed by animals is one of the major goals of research on captive animals. If these data sets yield significant allometric relationships, then the regression lines and associated statistics represent not only the general nature of water flux scaling, but also the variability within each taxonomic group. Moreover, if statistical comparisons between data sets show significant differences in spite of the variability within data sets, then conclusions drawn from these differences are robust, and probably reflect biological rather than methodological differences.

Some authors have used only one value, usually a mean of several measurements, per species when calculating allometric relationships of physiological or ecological variables. This approach yields a scaling relationship that is useful for several applications, including analysis of variation between species and characterization of fundamental properties within groups of species such as taxonomic families, orders, or classes. We used one or many values per species in the analyses herein, in order to reflect most effectively the literature available to date and to incorporate within-species variation into the regressions, for reasons given above. In many cases, the water flux values we used differed significantly within a species. These differences reflect variation due to age, sex, season, habitat, diet, and other biological and environmental variables, so the resulting allometric relationships should have greater ecological relevance than they would had we edited the data sets with a more narrow viewpoint. However, one data set (captive eutherians) contains so many values for some species (humans, and large domestic animals such as goats, sheep, and cattle) that the regression may be strongly biased toward large body masses. To test this, we recalculated the regression using species means only, and compared the two regressions (see below).

Each of the data sets was analyzed by least-squares regression of $\log_{10}$-transformed water flux rate (dependent variable) upon $\log_{10}$-transformed body mass (independent variable) (Dixon and Massey 1969). This procedure has been used in most previous analyses of scaling in animals (see recent reviews by Calder 1981 and 1984, Peters 1983, Schmidt-Nielsen 1984). We evaluated the question of whether water flux scales with body mass within data sets by using the F-statistic for significance of the regression. Then, we compared the regressions for captive and free-living animals within taxa by using analysis of covariance (ANCOVA) to test first for significant ($P < .05$) differences between slopes, and then, if slopes did not differ, for differences in intercepts, as recalculated on the basis of the common slope (Dunn and Clark 1974). Following these analyses, we looked within each data set for differences between subgroups based on diet and habitat categories. First, subgroups were tested for significance of the allometric regression by least-squares analyses. If regressions were statistically significant, ANCOVA was used to test for differences in slope. If slopes did not differ, the common slope was calculated, and intercepts were tested for differences. Wherever ANCOVA-generated common slopes are reported in regression equations given below, this is indicated. Otherwise, the equations and statistics shown were derived by the least-squares method. Estimates of variance for slope, intercept, and predicted log y values differ depending on whether ANCOVA or the least-squares method is used.

The equations for the regression lines have the form:
$$\log y = \log a + b \log x$$
where $\log y$ is $\log_{10}$ ml $H_2O$/day, $\log a$ is the intercept of the line and $a$ is the untransformed value of ml/day for a 1-g animal, $b$ is the slope of the line, and $\log x$ is $\log_{10}$ g (body mass). The statistics reported in Table A2 for each equation are: standard error of intercept ($SE_{\log a}$), confidence interval of intercept (95% $CI_{\log a}$), standard error of slope ($SE_b$), confidence interval of slope (95% $CI_b$), number of data points ($N$), coefficient of determination ($r^2$), probability value ($P$) for significance of regression from F-statistic, mean value of $\log_{10} x$ ($\overline{\log x}$), mean value of $\log_{10} y$ ($\overline{\log y}$), and the equation for calculating the 95% confidence interval (95% $CI_{\log y}$) of a predicted $\log y$ value ($\log y_p$) at any given $\log x$ value (Dunn and Clark 1974).

We compared allometric slopes derived herein ($b$) with previously published slopes ($b_{pp}$) for significant differences by using the equation $t = (b - b_{pp})/SE_b$, where $t$ is the t-statistic (Afifi and Azen 1979).

# SCALING IN MAJOR TAXONOMIC GROUPS

## EUTHERIAN MAMMALS

Daily water flux is highly correlated with body mass in captive eutherians ($P < 0.001$, $r^2 = 0.957$) as well as in free-ranging wild eutherians ($P < 0.001$, $r^2 = 0.894$; Fig. 1). (Complete regression statistics are summarized in Table A2.) The slopes of the regression lines differ significantly (ANCOVA: $P < 0.001$), and the lines cross at a body mass of about 300 g.

The slope of the line for captive eutherians (0.95, based on 562 points for 99 species) is significantly different from the slope of 0.82 for captive eutherians suggested by W.V. Macfarlane (1965), as well as the slope of 0.80 for 7 species of captive eutherians (Richmond, Langham, and Trujillo 1962), 0.78 for 12 species of captive rodents (Holleman and Dieterich 1973), and 0.82 for a data set representing 41 species of captive eutherians and marsupials combined (Nicol 1978). Our data set includes all the data used in these earlier analyses, as well as much new information, especially for large domestic eutherians such as sheep and cattle. Some sheep and cattle have been intentionally and unintentionally selected for high water flux, as this is often correlated with high wool, meat, or milk production. Also, high water flux may be a natural correlate of ruminant digestion (Maloiy et al. 1979). The only free-ranging wild ruminants studied (*Odocoileus hemionus* and *Bubalus bubalis*, Table A1) both have high water fluxes. In general, herbivores have higher water flux rates than do other mammals (see below). These ruminant data may account for the high slope for captive eutherians. To test whether the large numbers of data points for domestic ruminants biased the regression, from the viewpoint of its accurately representing between-species scaling (and ignoring the other purposes of allometric analyses explained above), we calculated a regression using a single mean data point for each

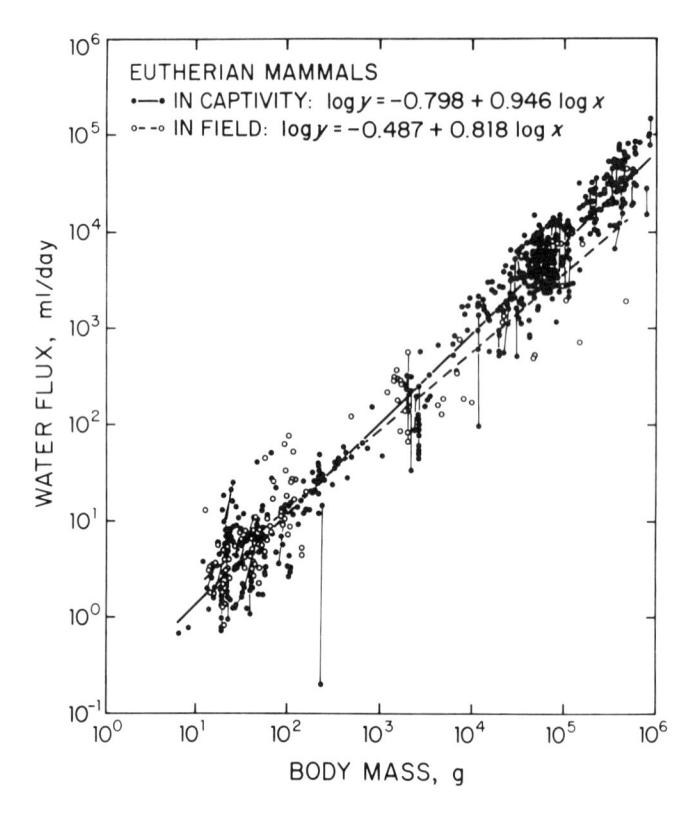

FIG. 1. Allometry of water flux rate among eutherian mammals, in captivity or domesticated (●), and free-living (o). Vertical and near-vertical lines illustrate intraspecific variability. Axes are logarithmic; lines represent least-squares regressions, and their equations are given at top of figure.

species ($N = 99$) and compared it via ANCOVA to the regression in Figure 1. The regressions do not differ significantly in either slope or intercept (both $P > 0.05$). The slope of 0.95 (Fig. 1) is close to, but still significantly different from, the slope of 0.88 for gravimetrically measured water intake rate in eutherians ranging from mice to elephants (Adolph 1949), as well as the slope of 0.91 derived from isotopic and other measurements of water flux in mammals (Streit 1982).

Variation of mean water flux values within a species, indicated by vertical or near-vertical lines connecting points in Figure 1, can be large. Intraspecific variability can be due to differences in ambient temperature, drinking-water availability, season, age, sex, or several other factors (Table A1). The highest water flux rates often exceed the lowest by 6-8 times, and 10-fold (order of magnitude) differences occur in some species. The largest range (longest vertical line in Fig. 1) yet measured is for a heterothermic mammal, the ground squirrel (*Spermophilus tridecemlineatus*; Table

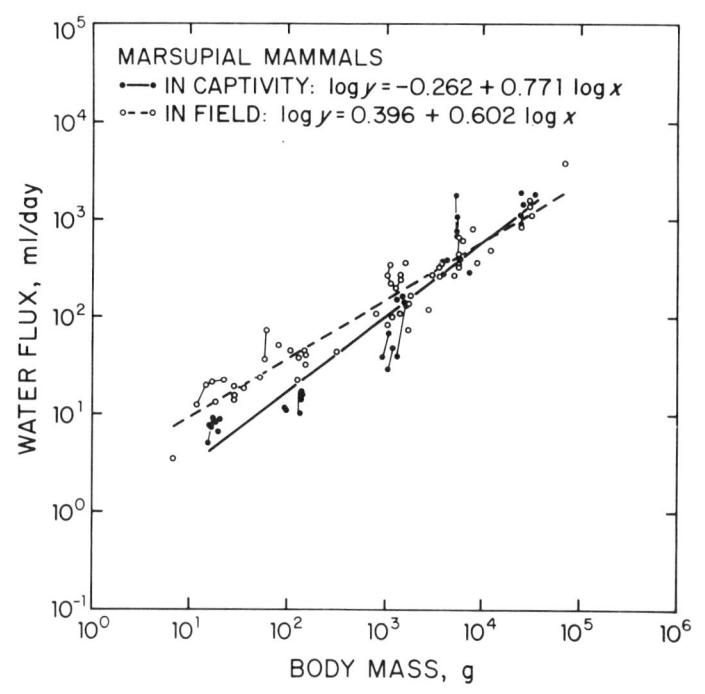

FIG. 2. Allometry of water flux rate among marsupial mammals (symbols as in Fig. 1).

A1), whose water flux while active in a cold environment was 72 times that during hibernation. Individual eutherian mammals are capable of an impressively wide range of variation in water flux. Unfortunately, most authors have not reported the rates of body mass change occurring in their animals while undergoing water flux measurements, so it is not yet possible to determine whether eutherians can maintain water balance over this range of water flux rates. The intraspecific variability evident in Figure 1 and Table A1 indicates that captive eutherians can be made to have water flux rates anywhere within a very large range of possible values.

Variation in water flux rates for free-living eutherians is also quite large (Fig. 1, Tables A1 and A2). However, this variation is due to natural rather than artificial causes. Interspecific differences can be due to differences in phylogeny, habitat, and diet, among other causes (see below). Within-species variation can result from ontogeny, season, gender, reproductive status, and weather conditions (Table A1). We know of no other allometric relationship for water flux in free-living eutherians with which to compare the line in Figure 1.

## MARSUPIAL MAMMALS

As with eutherians, water flux rate is highly correlated with body mass in captive marsupials ($P < 0.001$, $r^2 = 0.907$) as well as in free-living marsupials ($P < 0.001$, $r^2 = 0.917$, Fig. 2). These regressions have significantly different slopes ($P < 0.005$), indicating that water flux in marsupials scales differently when they are captive than when they are living in their natural habitats. In particular, small marsupials such as *Sminthopsis crassicaudata* and *Isoodon obesulus* (Table A1) have much higher water flux rates in the field than in captivity, and this accounts in part for the difference in slopes in Figure 2.

The slope for captive marsupials (0.77) does not differ significantly from the slope of 0.80 determined for captive macropodid marsupials (Denny and Dawson 1975) or the slope of 0.82 for captive eutherians and marsupials combined (Nicol 1978). No previously published allometric relationships for water flux in free-ranging marsupials are available for comparison with our regression.

The only monotreme mammal that has been studied is the highly aquatic platypus, and it has a very high water flux rate in the field (Table A1).

## BIRDS

Water flux in birds is also highly correlated with body mass, and the regressions for captive and free-ranging birds differ (Fig. 3). The slopes of the two relationships do not differ ($P < 0.25$), but the intercepts do ($P < 0.005$). The common slope is 0.694, and the common $r^2$ value is 0.834 for the two regressions.

Water flux in free-living birds is higher by about 50% than in captive birds, regardless of body mass. We suspect this is due in part to greater activity and food consumption in wild vs. caged birds. Most field studies were done on birds that were breeding (because of the difficulty of repeatedly recapturing birds during nonbreeding seasons), and breeding birds might be expected to have higher food and water intake rates than nonbreeding birds (Whittow 1986). The largest variation in daily water flux within species, as indicated in Figure 3, reflects the differences between birds on a nest (incubating or brooding) and off the nest (primarily foraging) while breeding.

Our allometric slope for captive birds (0.69, based on 74 points for 28 species) does not differ significantly from the earlier slopes of 0.69 (5 points for 4 species; Ohmart et al. 1970), 0.71 (13 species; Degen, Pinshow, and Alkon 1982), 0.75 (19 species; Pinshow et al. 1983), 0.660 (52 points for 22 species; Booth 1987), and 0.73 (10 species; Hughes et al. 1987), but 0.69 does differ significantly ($P < 0.01$) from the slope of 0.596 calculated by Degen, Pinshow, and Alkon (1982) for 13 species excluding Pekin ducks,

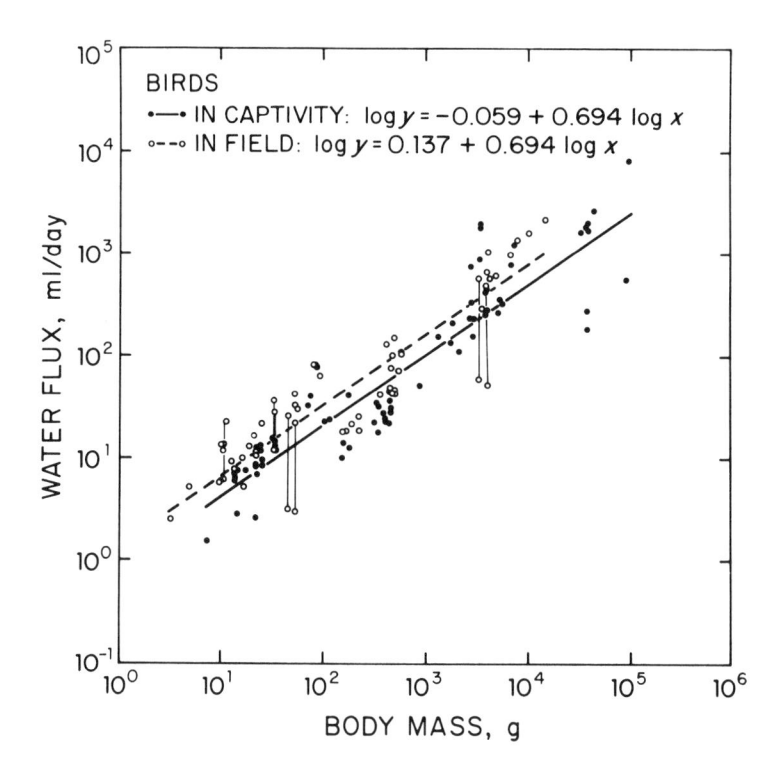

FIG. 3. Allometry of water flux rate among birds (symbols as in Fig. 1).

and from the slope of 0.75 ($P < 0.05$) for captive birds (20 points for 10 species; Walter and Hughes 1978), as well as the slope of 0.75 ($P < 0.05$) for birds with salt glands (10 species; Hughes et al. 1987). Our regression for water flux scaling in free-living birds (slope = 0.69, 62 points for 27 species) does not differ significantly from the slope of 0.662 for birds in the field (21 points for 11 species; Booth 1987).

## REPTILES

Water flux increases with increasing body mass in reptiles (Fig. 4), but the allometric relationships for captive and free-living reptiles differ. The slopes do not differ ($P > 0.05$) but the intercepts do ($P < 0.001$; common $r^2 = 0.648$), with captive reptiles having water fluxes about 3 times higher than those in the field. The primary reason for this difference is that most data for captive reptiles are for aquatic forms measured while in water, while most field data are for terrestrial species (Table A1). In the three species which were studied in captivity and in the field, one (*Lacerta viridis*) had a

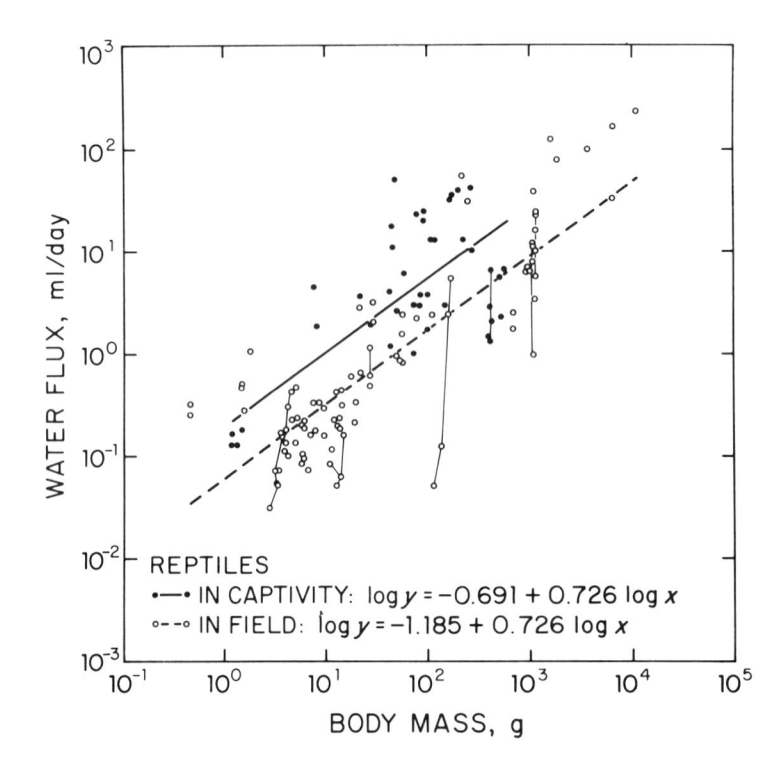

FIG. 4. Allometry of water flux rate among reptiles (symbols as in Fig. 1).

lower water influx in captivity than in the field, and the two others (*Sauromalus* and *Gopherus*) had water fluxes in captivity that were intermediate to field values.

The scaling of water flux in captive reptiles has not been reported previously. However, Minnich (1979) derived a slope of 0.84 for free-living lizards, which differs significantly ($P < 0.02$) from the slope of 0.73 in Figure 4. The slope of 0.73 also differs significantly from the slope of 0.91 for arid-habitat reptiles (Nagy 1982), but does not differ from the slope of 0.66 for tropical reptiles (Nagy 1982). Habitat differences are discussed in more detail below.

Variability in water flux within species of reptiles can be quite large. The greatest variation measured to date (longest vertical line in Fig. 4) occurred in chuckwalla lizards (*Sauromalus obesus*), whose water flux rate in spring was 73 times that during winter hibernation (Table A1). This is similar to the 72-fold variation measured in hibernating vs. active ground squirrels. Hibernation and estivation are phenomena that greatly reduce water requirements as well as energy requirements.

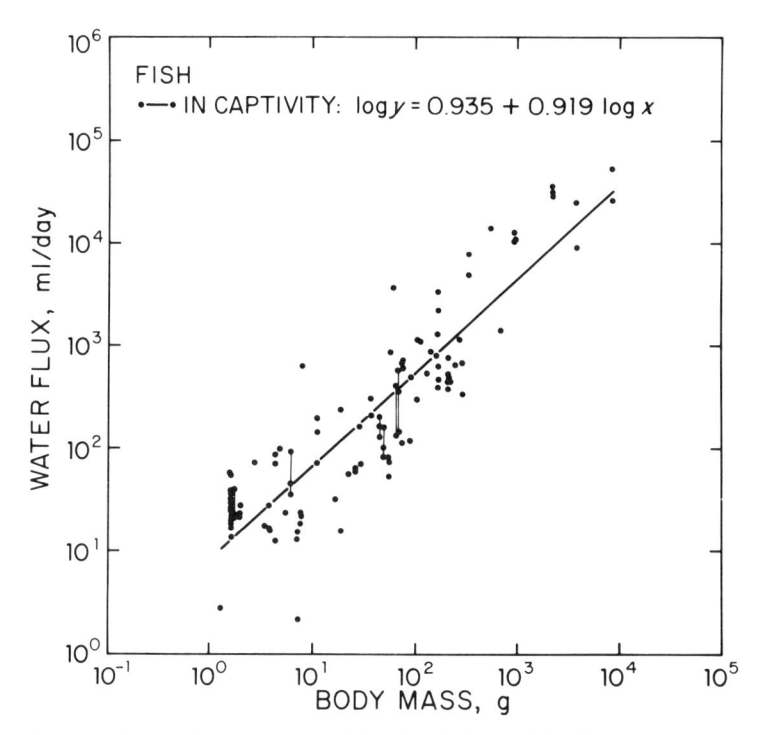

FIG. 5. Allometry of water flux rate among fishes (symbols as in Fig. 1).

## AMPHIBIANS

Adequate numbers of water flux measurements are not yet available to test for a significant correlation with body mass among adult amphibians (Table A1). Water-breathing larvae have flux rates similar to those in fish, but the air-breathing adults have flux rates near those of endothermic vertebrates. A correlation does exist within aquatic larval amphibians (see below). Only one species of amphibian has been studied in the field (Table A1), and field studies of amphibian water relations would provide valuable and interesting information.

## FISHES

Water flux is strongly correlated ($P < 0.001, r^2 = 0.821$) with body mass in captive fish (Fig. 5). No field studies have been done on fish. This is understandable because their high rates of water flux dictate very short measurement intervals (minutes or a very few hours), which render field studies difficult and of dubious value, because of probable large artifacts from the effects of investigator disturbance.

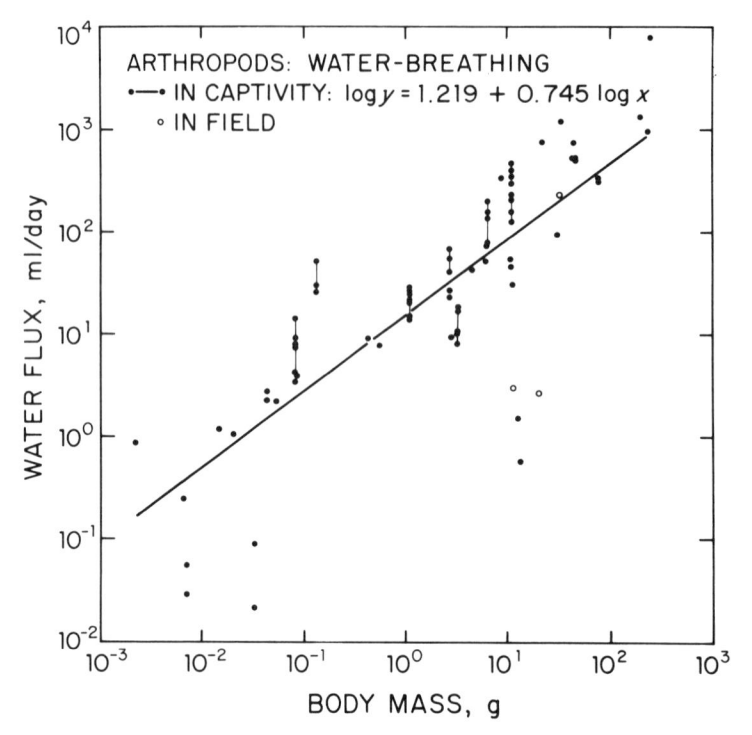

FIG. 6. Allometry of water flux rate among water-breathing arthropods (symbols as in Fig. 1).

The slope of 0.92 in Figure 5 does not differ significantly from that of 0.88, which is the average of within-species slopes for 5 species of teleosts (Evans 1969b). Variation of water flux within individual fish species (Fig. 5, Table A1) occurs mainly in response to differences in temperature (positive relationship) and salinity (discussed below).

## ARTHROPODS

### Water-breathing

Although the data are quite variable (Fig. 6), water flux is significantly correlated ($P < 0.001$, $r^2 = 0.642$) with body mass in water-breathing arthropods in captivity. Only two field studies have been done on water-breathing arthropods (Table A1). These data show no significant regression, so they were combined with captive animal data to calculate the line in Figure 6. Within-species variation in water flux is associated mostly with differences in salinity, with water flux tending to increase at higher

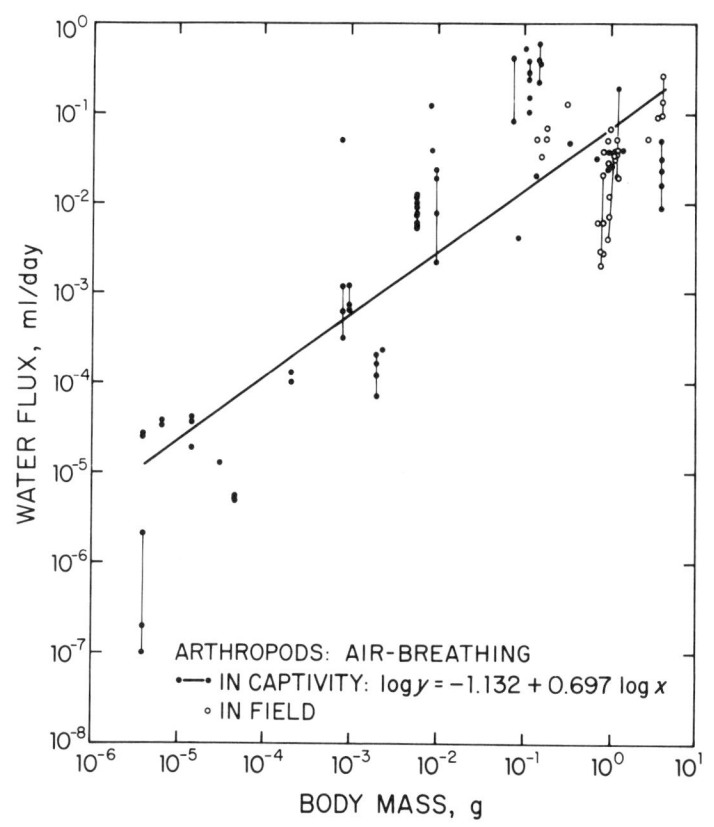

FIG. 7. Allometry of water flux rate among air-breathing arthropods (symbols as in Fig. 1).

salinities in most species (Table A1). No other water flux regressions for arthropods are available for comparison.

## Air-breathing

Water flux rate did not correlate significantly with body mass for the data on free-living, air-breathing arthropods (Fig. 7), so these data were combined with the data for captive animals. Water flux rate increases with increasing body mass in air-breathing arthropods ($P < 0.001$, $r^2 = 0.689$).

There is much variation in water flux rate, within species as well as between species of air-breathing arthropods (Fig. 7). The coefficients of determination ($r^2$) are generally lower, and the 95% confidence intervals of the regression parameters $a$ and

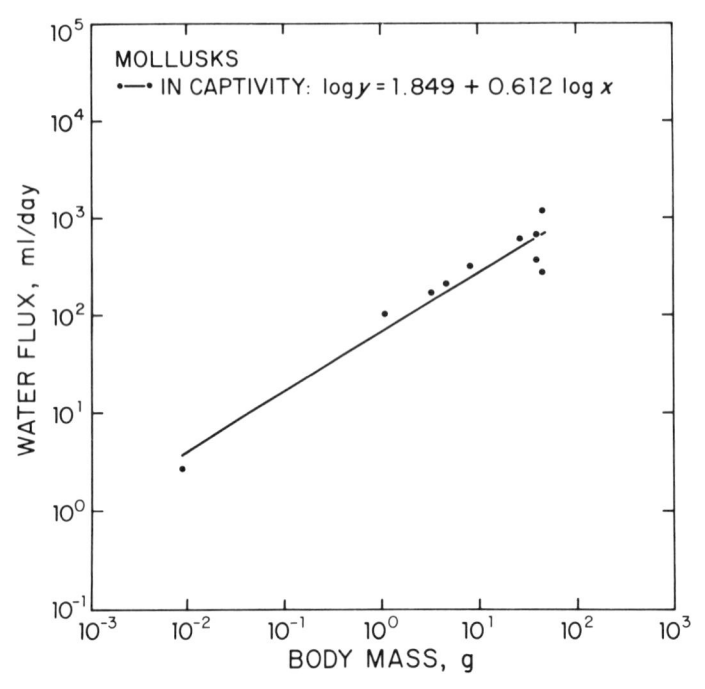

FIG. 8. Allometry of water flux rate among mollusks (symbols as in Fig. 1).

$b$ are generally larger for arthropods than for vertebrate taxa (Table A2). This indicates that the allometric relationships for arthropods are less reliable as predictive models for water flux than are those for vertebrates. However, this variability provides the hint that a rich diversity of water balance adaptations exists among arthropods. More than any vertebrate group, air-breathing arthropods should yield exciting results upon investigation in the field.

## MOLLUSKS

Water flux in captive aquatic mollusks is highly correlated ($P < 0.001, r^2 = 0.929$) with soft body mass (Fig. 8). No field studies of mollusks have yet been done, and no other allometric regression or water flux in mollusks is available for comparison.

## ANNELIDS

Only three species of polychaete worms have been studied to date (Table A1). These data do not yield a significant allometric regression ($P > 0.25$), probably because the range of body masses is narrow.

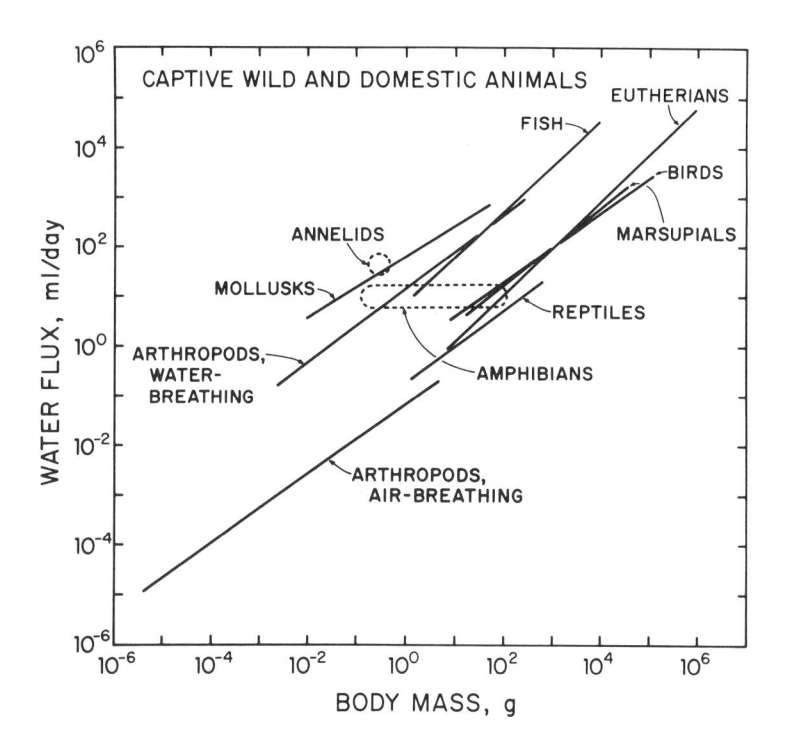

FIG. 9. Comparison of allometric regressions among major taxonomic groups. Measurements were made on wild animals in captivity or on domesticated animals.

## SINGLE CELLS

Water flux has been estimated for human erythrocytes, vertebrate muscle cells, and fish eggs (Table A1). These data do not show a significant correlation ($P > 0.10$) between log water flux rate and log mass.

## SUMMARY

To summarize this section, water flux rate is highly correlated with body mass (logarithmically transformed data) in all animal groups for which sample size is substantial. Water flux scales differently in the field than it does in captive animals, suggesting that quantitative ecological conclusions based on studies of water flux rates of captive animals may be misleading. Variation is especially large among arthropods, indicating that this group contains a comparatively great diversity of water balance adaptations.

# COMPARATIVE SCALING

## IN CAPTIVITY

Water-breathing animals (mollusks, fish, water-breathing arthropods) have the highest rates of water flux; air-breathing endothermic animals (mammals and birds) have lower rates; and air-breathing ectothermic animals (reptiles and air-breathing arthropods) have the lowest water fluxes (Fig. 9). Annelid worms have rates near the water-breathing groups. Larval amphibians are also close to other water-breathers, while the air-breathing adult amphibians overlap with small mammals and birds (Fig. 9). The regression line for captive reptiles is higher than the line for air-breathing arthropods, and approaches the lower end of the eutherian line. However, most captive reptiles studied were aquatic, and the few terrestrial reptiles studied have water fluxes that would fall near the air-breathing arthropod line, if it were extended.

The slopes for the three groups of water-breathing animals all differ from each other (ANCOVA, $P < 0.025$ for the three regressions combined, and $P < 0.025$ comparing water-breathing arthropods with fish). Among air-breathing endotherms, the slope for eutherians is different ($P < 0.001$) from those for marsupials and birds, but the marsupial and bird regressions have slopes ($P > 0.25$) and intercepts ($P > 0.75$) that do not differ significantly. Nicol (1978), analyzing a smaller data set, found no significant difference between eutherians and marsupials. The slopes for reptiles and air-breathing arthropods do not differ significantly ($P > 0.25$), but reptiles have a significantly different intercept ($P < 0.025$).

## IN THE FIELD

Data sets that are adequate for allometric analysis are available only for eutherians, marsupials, birds, and reptiles (Fig. 10). As with captive endotherms, free-living

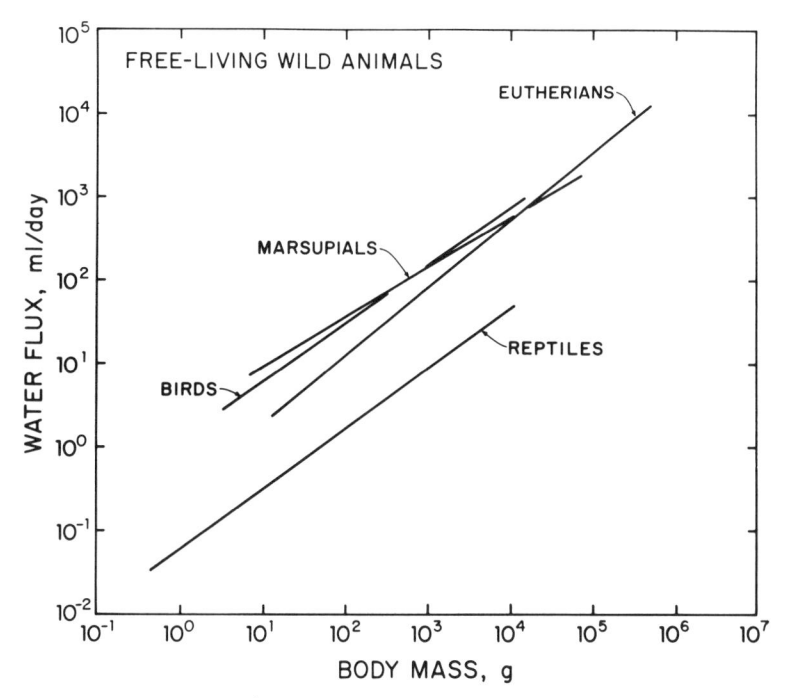

FIG. 10. Comparison of allometric regressions for water flux among terrestrial vertebrate taxa. Measurements were made on wild animals living in their natural habitats.

endotherms have generally similar water flux rates, and these are much higher than those in similar-sized reptiles. Among the endotherms, eutherians have a significantly different slope ($P < 0.001$) from marsupials and birds, and the latter have statistically indistinguishable slopes ($P > 0.25$) and intercepts ($P > 0.25$).

The slope for reptiles does not differ from those for eutherians ($P > 0.25$) and birds ($P > 0.10$), but it is different from the marsupial slope ($P < 0.025$).

Thus, scaling differences among major taxonomic groups exist within the three categories of water-breathers, air-breathing endotherms, and air-breathing ectotherms. This indicates that a list of factors contributing to variation in water flux scaling should include (1) nature of body temperature maintenance (endotherm or ectotherm) and (2) gas exchange medium (air or water), as well as (3) taxonomic affiliation. In the next section, we explore the effects of other factors on water flux scaling.

# DIET AND HABITAT EFFECTS

We used the scaling relationships for major taxonomic groups (above) to predict water flux rate for each species listed in Table A1 (for those groups having significant regressions). Then, we expressed actual water flux as a percentage of the predicted value. Examination of these values (Table A1) reveals much variation, some of which may be due to diet and environmental parameters.

Most terrestrial, air-breathing animals obtain most of their water from the food they eat and from drinking liquid water. Various diets differ in the amount of water they contain: green plant matter typically is quite succulent, animal food usually contains somewhat less water, and mature seeds contain little water (Edney and Nagy 1976). We used ANCOVA to test for scaling differences by diet category within the data sets for free-living terrestrial vertebrates. Similarly, the availability of drinking water in terrestrial habitats varies considerably, and habitat aridity may have a substantial effect on water flux in resident animals. We examined the data sets for free-living terrestrial vertebrates for scaling differences between desert-dwelling and nondesert-inhabiting species. The data for captive terrestrial vertebrates were not tested for diet or habitat differences, because the wide variety of conditions under which these animals were studied makes interpretation of any differences very complicated and of dubious usefulness (Nicol 1978; Degen, Pinshow, et al. 1981). Among captive water-breathing animals, we tested for scaling differences due to salinity of the medium (sea water vs. fresh water) and for differences associated with different osmoregulatory mechanisms (osteichthyes vs. chrondrichthyes). Also, in an attempt to account for some of the variation noted above, we tested the data set for air-breathing arthropods for differences due to habitat.

## Diet and Habitat Effects

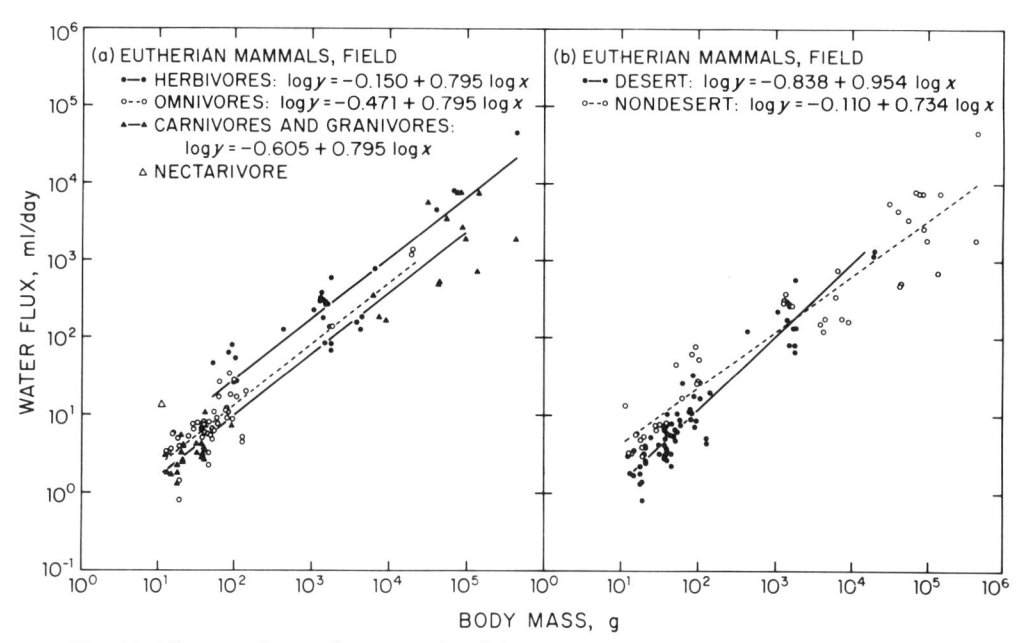

FIG. 11. Allometry of water flux among free-living eutherian mammals according to differences in (a) diet and (b) habitat.

## EUTHERIANS IN THE FIELD

Free-living herbivorous eutherians ($N = 28$) have water fluxes that are almost 3 times higher than those in carnivorous and granivorous eutherians (combined; $N = 39$), and omnivorous eutherians ($N = 47$) have intermediate water fluxes (Fig. 11a). The slopes of these regressions do not differ ($P > 0.50$), but the intercepts do ($P < 0.001$). (Carnivores and granivores do not differ, and are combined.) The only nectarivorous eutherian that has been studied (the bat *Anoura caudifer*, Table A1) has a very high water flux rate.

The high water fluxes in herbivorous mammals are expected on the basis of the high water content of plant tissues, coupled with the comparatively low digestibility of plant diets by mammals. The difficulty of breaking down cell-wall constituents means that many herbivorous animals must consume more dry matter (relative to carnivorous or omnivorous animals) to obtain a given amount of food energy (Shoemaker and Nagy 1977). Thus, herbivores tend to consume more dietary water than nonherbivores. The lack of difference in water flux scaling between carnivores and granivores is surprising, and may result from granivorous mammals' obtaining more water from their environment than expected for a diet of dry seeds alone. This

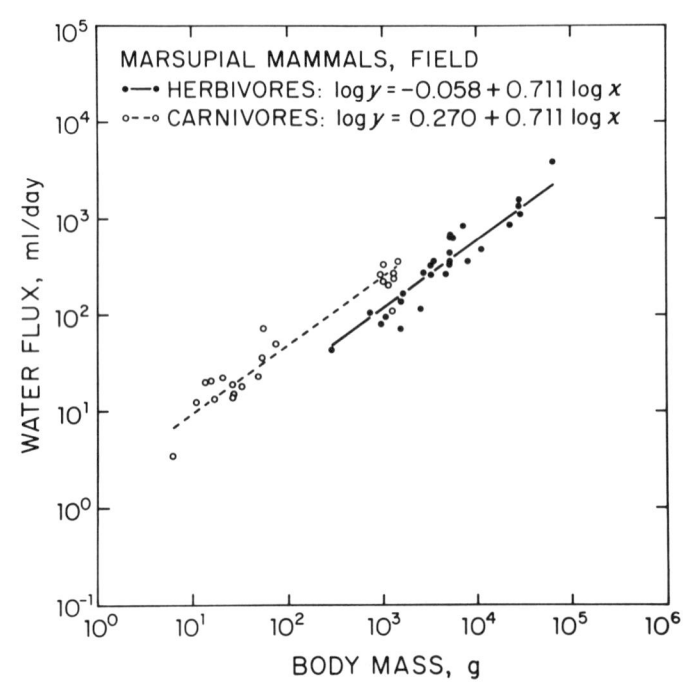

FIG. 12. Allometry of water flux among herbivorous and carnivorous marsupials in the field.

possibility can be evaluated more rigorously with the water economy index (WEI: see below).

Water flux scales differently in desert eutherians ($N = 48$) than in nondesert forms ($N = 78$) in the field (Fig. 11b): the regression slopes differ significantly ($P < 0.001$). These relationships indicate that small ($<2$ kg) desert eutherians typically have lower water fluxes than nondesert eutherians of the same body mass. The comparatively low water fluxes of small desert eutherians, which are primarily omnivorous and granivorous rodents (Table A1), are expected in view of their diet and arid habitat. This expectation is based on the reasonable assumption that desert eutherians are adapted to their habitat through water conservation mechanisms that reduce their water requirements below those of nondesert relatives. However, it is not necessary to have low water requirements to live in deserts. An animal may possess instead the ability to obtain much water from its desert habitat, and survive well despite a mediocre or even extravagant water requirement. Some of the larger desert mammals depicted in Fig. 11b apparently possess behavioral traits (such as using scattered oases for drinking, selection of succulent plant foods) that enhance their water intake rates,

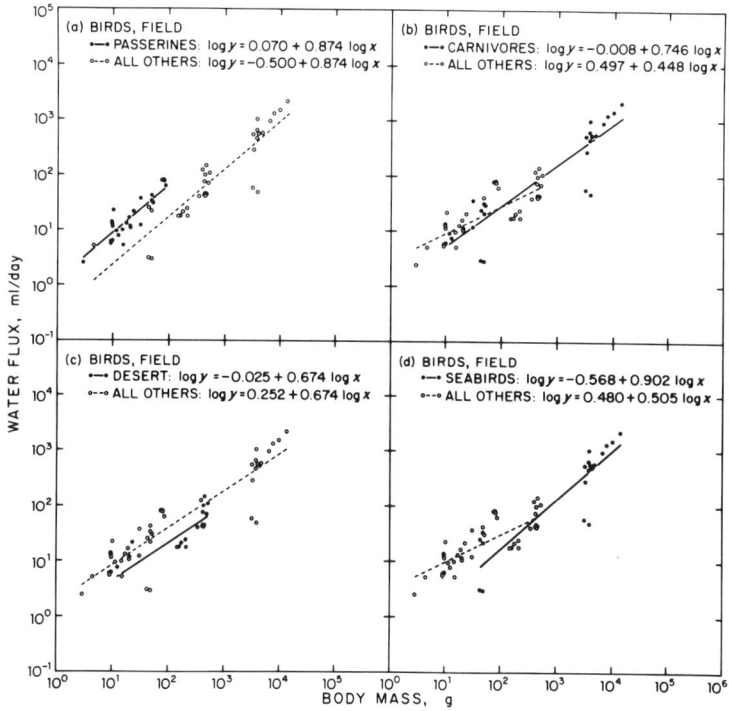

FIG. 13. Allometry of water flux in free-living birds in relation to (a) taxon, (b) diet, and (c) and (d) habitat.

and yield the comparatively high slope (0.954) for desert eutherians. It would be interesting to measure field water fluxes of small herbivorous or insectivorous desert eutherians, such as wood rats (*Neotoma*), grasshopper mice (*Onychomys*), and bats (e.g., *Pipistrellus*) to evaluate their modes of water balance maintenance.

## MARSUPIALS IN THE FIELD

Surprisingly, herbivorous marsupials ($N = 28$) have water flux rates that are less than half those of carnivorous marsupials ($N = 23$; Fig. 12). The slopes do not differ ($P > 0.05$), but the intercept for herbivores is lower than for carnivores ($P < 0.001$), with $r^2 = 0.907$ and $N = 23$. The herbivorous marsupials include the arboreal ringtail possums, greater gliders, and koalas, which eat eucalyptus leaves and other sclerophylous vegetation, and macropods that consume grass, which may contain less water than more succulent vegetation. Most of the marsupial carnivores that have been studied belong to the family Dasyuridae, which are active animals that live in moist habitats and may have more frequent access to drinking water than do the herbivores. Only a few desert marsupials have been studied in the field, and this lack of data precludes allometric analysis.

## BIRDS IN THE FIELD

Birds in the order Passeriformes are known to have higher basal and field metabolic rates than do other birds (Lasiewski and Dawson 1967; Nagy 1987). An ANCOVA comparison of water flux allometry shows that passerines ($N = 26$) also have higher water fluxes than do other birds (Fig. 13a): the slopes do not differ ($P > 0.25$), but the intercepts do ($P < 0.001$). These regressions indicate that passerine birds take in 3.7 times more water each day than do other birds. Part of this difference is due to the nearly twofold higher field metabolic rate in passerines (Nagy 1987), and the remainder is probably due to differences in diet and drinking behavior.

Diet also has an effect on water flux scaling in birds (Fig. 13b). The slope for carnivorous birds ($N = 18$) is significantly different ($P < 0.025$) from that for other species. Water flux scaling in birds is further complicated by habitat differences (Fig. 13c and d). Desert birds ($N = 18$) have significantly lower water fluxes than nondesert birds [slopes do not differ ($P > 0.25$), but intercepts do ($P < 0.005$)]. These regressions indicate that water fluxes in desert birds average about half those of nondesert birds. Similarly, water flux in seabirds ($N = 18$) scales differently from that in other birds [slopes differ ($P < 0.001$)].

The data presently available for birds show substantial and highly significant differences in water flux scaling in relation to taxon, diet, and habitat. It will be interesting to determine whether, for example, very large or very small desert birds conform to predictions, or how these variables interact within a species, such as a small desert passerine bird that is carnivorous (e.g., shrikes). Many of the bird species that are carnivorous (Fig. 13b) are also seabirds (Fig. 13d). The apparent coupling between these two natural history traits could be tested by studying hawks, which are large nonseabird carnivores.

## REPTILES IN THE FIELD

Water flux does not scale differently between herbivorous and carnivorous reptiles ($P > 0.50$ for slope and $P > 0.05$ for intercept). Previous analysis of a subset of these data also indicated no dietary differences, but this conclusion was tempered by the observation that the herbivorous reptiles did not drink water, but most nonherbivorous reptiles had access to drinking water (Nagy 1982). Field measurements on carnivorous reptiles without drinking water still remain to be done before this question can be resolved adequately.

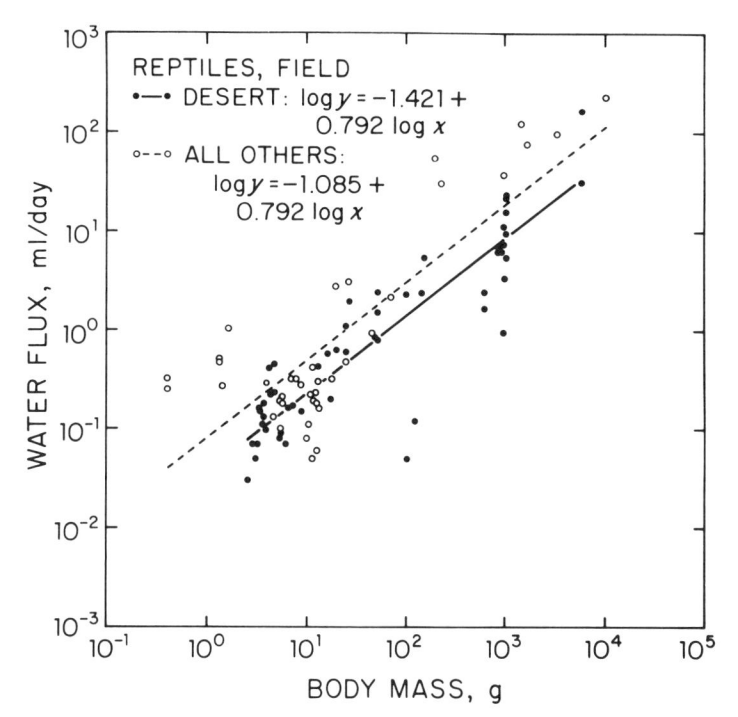

FIG. 14. Allometry of water flux in free-living desert and nondesert reptiles.

Desert-dwelling reptiles (Fig. 14) have water fluxes averaging about half those of other reptiles [slopes do not differ ($P > 0.25$), but intercepts do ($P < 0.001$)]. Nagy (1982), using a subset of the data used herein, reported scaling slopes of 0.91 for reptiles living in arid and semi-arid habitats, and 0.66 for tropical and subtropical reptiles. Both of these slopes differ significantly ($P < 0.02$) from the common slope of 0.792 calculated herein. Minnich (1979), using a smaller data set, determined a slope of 0.84 for lizards, which does not differ significantly ($P > 0.20$) from the common slope of 0.792. The equation in Figure 14 is based on all data presently available, and therefore it should be the most reliable.

## AMPHIBIANS IN CAPTIVITY

Although no significant regression exists among the amphibian data as a whole, water flux is significantly ($P < 0.001$) correlated with body mass in aquatic larval (water-breathing) amphibians. The relationship is:
$$\log y = 1.765 + 0.980 \log x$$
($r^2 = 0.958$, $N = 11$). This line is significantly higher than the line for fish [slopes do

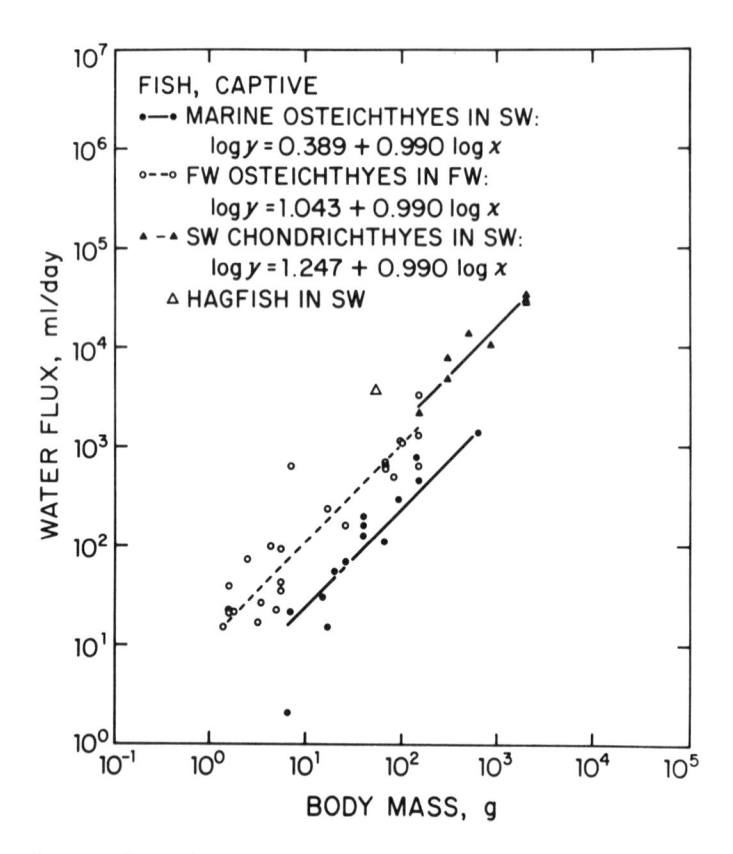

FIG. 15. Allometry of water flux among osteichthian and chondrichthian fishes in sea water (SW) and fresh water (FW) in captivity.

not differ ($P > 0.75$) but intercepts do ($P < 0.001$)]. Water-breathing amphibians have water fluxes about 6 times higher than fish having the same body mass. Aquatic amphibians exchange water across gills and through their highly permeable skins (Alvarado 1979), whereas fishes have relatively impermeable integuments (Evans 1979).

## FISHES IN CAPTIVITY

Within the fishes, there are significant differences according to taxonomic class and salinity of habitat [sea water (SW) vs. fresh water (FW)]. ANCOVA of captive marine osteichthyes measured in SW ($N = 14$), captive FW osteichthyes in FW ($N = 26$), and captive marine chondrichthyes in SW ($N = 11$) indicates that the slopes do not differ ($P > 0.10$), but the intercepts do ($P < 0.001$). The differences in intercepts (Fig.

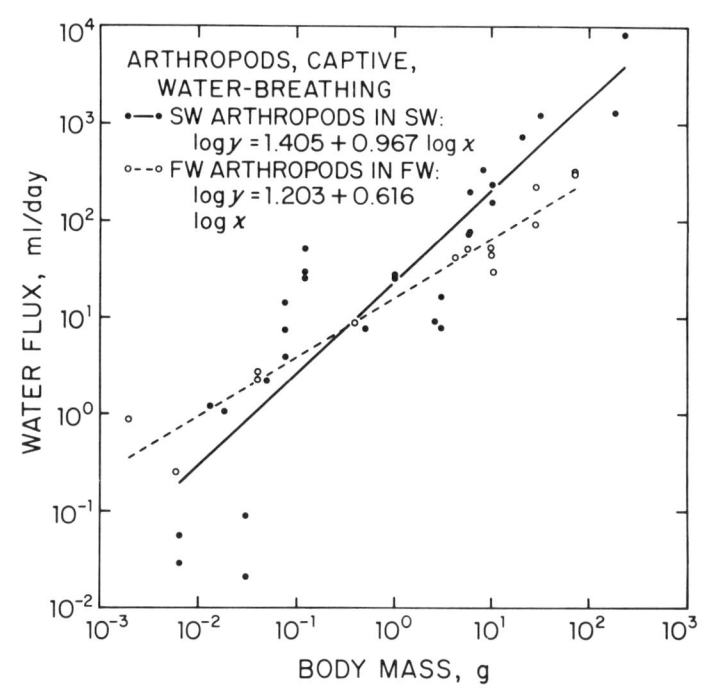

FIG. 16. Influence of water salinity (SW = sea water, FW = fresh water) on allometry of water flux in water-breathing arthropods in the laboratory.

15) indicate that average water flux in SW osteichthyes living in SW is about 14% that of SW chondrichthyes in SW, and FW osteichthyes in FW average about 63% that of SW chondrichthyes in SW. Thus, FW bony fish turn over water about 4.5 times faster than their SW counterparts. This difference is about twice as great as the twofold difference found in 5 species each of marine and freshwater teleosts (Evans 1969b). The differences in osmoregulatory modes used by FW teleosts, SW teleosts, and SW elasmobranchs are well known (Krogh 1939, Evans 1979), and the allometric relationships reported here conform well to expectations from current models. In general, the smaller is the osmotic gradient between the body fluids of fish and the environment, the greater the permeability of the fish should be to water, and the greater should be the unidirectional flux of water (Evans 1979). Among teleosts, FW forms have low osmotic gradients and high water fluxes, while SW forms have high gradients and low fluxes, in conformity with expectations. Moreover, water fluxes are quite high in marine elasmobranchs, which maintain body fluids that are isosmotic with or slightly hyperosmotic to SW by retaining urea. Interestingly, the Pacific hagfish (*Eptatretus stoutii*, Table A1), which remains isosmotic with SW without the use of

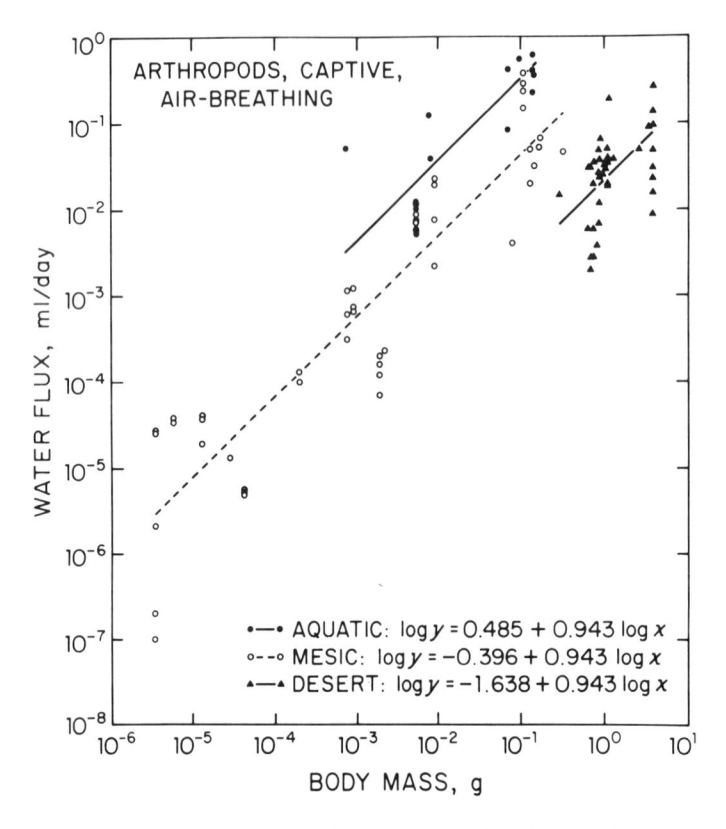

FIG. 17. Allometry of water flux in air-breathing arthropods in relation to habitat.

organic molecules such as urea but by retention of certain electrolytes, has an extremely high water flux rate, more than 4 times that expected for a marine elasmobranch of equivalent body mass. This general pattern holds within species for most anadromous and euryhaline fishes that have been studied: water flux in FW is usually greater than that in SW (Table A1). Results for the anadromous eel *Anguilla anguilla* are equivocal. Because isotopically determined water flux rates in water-breathing animals represent primarily the diffusional exchange of water across respiratory surfaces, these differences between SW and FW fish mostly reflect differences in the water permeability of gills (Evans 1969b).

## ARTHROPODS IN CAPTIVITY

Among the water-breathing arthropods, water flux scales differently in captive SW ($N = 28$) vs. FW ($N = 14$) species [slopes differ ($P < 0.005$)] (Fig. 16). There are also habitat differences among air-breathing arthropods in captivity (Fig. 17). ANCOVA indicates that the regressions for air-breathing aquatic species measured while in water ($N = 14$), mesic species measured in air ($N = 48$), and desert species measured in air ($N = 39$) all have similar slopes ($P > 0.25$), but the intercepts differ ($P < 0.001$). Mesic arthropods have water fluxes averaging 9 times higher than those of desert arthropods, and aquatic, air-breathing arthropods have water fluxes averaging 100 times higher than those in desert species. For comparison, a typical 1-g water-breathing arthropod in FW has a water flux rate about 7 times higher than that of a 1-g air-breathing arthropod swimming in or on FW, and about 700 times that of a 1-g desert arthropod. Thus, some of the extreme variability in water flux rate among arthropods can be correlated with differences in habitat, and with the gas exchange medium used (water or air). Some of this variation in water flux may also be due to differences in metabolic rate (see below).

# PREDICTING WATER FLUX RATE

The allometric equations above are useful for characterizing and comparing water fluxes in various groups of animals and for various environmental conditions. These equations may also be used to predict, from knowledge of body mass, taxon, diet, and habitat, water flux rates for species that have not yet been studied. As such, these empirically derived regressions can serve as "standards," against which one may compare data on additional species in order to reveal adaptations that enhance water balance. The exponential forms of these equations are the easiest to use for making predictions and for interpreting the intercept values (which show water flux rate for a 1-g animal). These equations are summarized in Table 1, along with the 95% confidence interval of the predicted water flux rate for an animal having a body mass equal to the antilog of mean $\log x$ for the regression. These 95% confidence intervals, calculated using the equations in Table A2, are for a predicted $\log y$ value, and they are much larger than the 95% confidence intervals for the regression. Many earlier publications dealing with allometry have reported confidence intervals for the regression only, erroneously implying that confidence intervals for the regression are also appropriate for values predicted from that regression. The confidence intervals for the prediction are smallest at mean $\log x$, and become larger as body mass increases or decreases from mean $\log x$. The equations in Table A2 can be used to calculate 95% CI for a predicted water flux rate at any given body mass.

TABLE 1. Summary of allometric equations for water flux rate in animals. The equations have the form $y = ax^b$, where $y$ is water flux rate (in ml/day) and $x$ is body mass (in g).

| Group | a | b | 95% CI of predicted $y$ as % of $y_p$ [1] | Equation |
|---|---|---|---|---|
| EUTHERIAN MAMMALS | | | | |
|   In captivity | 0.159 | 0.946 | −75 to +300% | (1) |
|   In field | 0.326 | 0.818 | −80 to +400% | (2) |
|     Herbivores | 0.708 | 0.795* | −74 to +280% | (3) |
|     Omnivores | 0.338 | 0.795* | −74 to +280% | (4) |
|     Carnivores and granivores | 0.248 | 0.795* | −74 to +280% | (5) |
|     Desert eutherians | 0.145 | 0.954 | −84 to +510% | (6) |
| MARSUPIAL MAMMALS | | | | |
|   In captivity | 0.547 | 0.771 | −72 to +260% | (7) |
|   In field | 2.488 | 0.602 | −59 to +140% | (8) |
|     Herbivores | 0.874 | 0.711* | −53 to +110% | (9) |
|     Carnivores | 1.862 | 0.711* | −54 to +120% | (10) |
| BIRDS | | | | |
|   In captivity | 0.874 | 0.694* | −78 to +350% | (11) |
|   In field | 1.369 | 0.694* | −78 to +350% | (12) |
|     Passerines | 1.180 | 0.874* | −73 to +270% | (13) |
|     Carnivores | 0.981 | 0.746 | −86 to +630% | (14) |
|     Desert birds | 0.944 | 0.676* | −77 to +340% | (15) |
|     Seabirds | 0.270 | 0.902 | −88 to +730% | (16) |
| REPTILES | | | | |
|   In captivity | 0.204 | 0.726* | −91 to +1000% | (17) |
|   In field | 0.065 | 0.726* | −91 to +990% | (18) |
|     Desert reptiles | 0.038 | 0.792* | −89 to +800% | (19) |
| AMPHIBIANS | | | | |
|   In captivity: water-breathing larvae | 58.2 | 0.980 | −39 to +64% | (20) |
| FISHES | | | | |
|   In captivity [2] | 8.61 | 0.919 | −86 to +620% | (21) |
|     SW osteichthyes in SW | 2.45 | 0.990* | −74 to +290% | (22) |
|     FW osteichthyes in FW | 11.0 | 0.990* | −74 to +280% | (23) |
|     SW chondrichthyes in SW | 17.6 | 0.990* | −75 to +300% | (24) |
| ARTHROPODS | | | | |
|   In captivity | | | | |
|     Water-breathers [2] | 16.6 | 0.745 | −95 to +1860% | (25) |
|       SW arthropods in SW | 25.4 | 0.967 | −96 to +2150% | (26) |
|       FW arthropods in FW | 16.0 | 0.616 | −70 to +240% | (27) |
|     Air-breathers | 0.074 | 0.697 | −98 to +4720% | (28) |
|       Aquatic arthropods in water | 3.05 | 0.943* | −93 to +1420% | (29) |
|       Mesic arthropods in air | 0.402 | 0.943* | −93 to +1270% | (30) |
|       Desert arthropods in air | 0.023 | 0.943* | −93 to +1330% | (31) |
| MOLLUSKS | | | | |
|   In captivity: aquatic | 70.7 | 0.612 | −68 to +220% | (32) |

[1] Calculated at mean log $x$ for the regression (see text for details). The confidence intervals are symmetrical in log form but asymmetrical in the antilog form shown here.

[2] SW = seawater; FW = freshwater.

* Common slope from ANCOVA.

# WATER ECONOMY INDEX

The adjustments that terrestrial animals have made to maintain water balance in their water-scarce habitats (relative to aquatic habitats) should be reflected in the rates at which they process water. The hierarchy of water flux that emerges from Figures 9 and 10 is water-breathers > endothermic air-breathers > ectothermic air-breathers. Within the air-breathing groups, desert animals have consistently lower water fluxes than nondesert animals. These hierarchies suggest the hypotheses that reptiles and air-breathing arthropods possess more effective water-conserving adaptations (physiological, behavioral, morphological) than do mammals and birds; and that desert mammals, birds, and reptiles conserve water more effectively than do their nondesert relatives. An alternative explanation is that these differences in water flux rates merely reflect the well-known differences in metabolic rates between these groups of animals. Reptiles and terrestrial arthropods have much lower metabolic rates than do endotherms, and many desert animals have somewhat lower metabolic rates than related nondesert animals, both in the laboratory and in the field (Bartholomew 1982, Schmidt-Nielsen 1983, Nagy 1987). In other words, do some animals use less water each day than others because they are better adapted to arid conditions, or simply because they use less energy each day?

The "water economy index" (WEI) expresses water flux relative to energy metabolism, and WEI values can be used to test the above hypotheses. The WEI is the ratio of daily water flux rate (ml/day) to daily metabolic rate (kJ/day), so WEI has the units of milliliters of water processed per kilojoule of energy metabolized. Because WEI indicates the amount of water used per unit of "living" an animal does, it is independent of the rate at which an animal uses energy, and can be used to compare directly the water economy of animals in different taxa and of different body masses. WEI is similar to the "water use efficiency" ratio used by plant physiologists to evaluate the amount of $CO_2$ taken up per unit of water transpired by plants (Nobel 1980 and 1983, Ehleringer et al. 1985).

Evaluation of water economy of animals in relation to their energy metabolism has been suggested by numerous authors (Macfarlane, Howard, et al. 1971, Yousef et al. 1974, Deavers and Hudson 1977, Withers et al. 1980, Nagy 1982, Hinds and MacMillen 1985). Most earlier analyses have been done on captive or domestic animals, and some have incorporated basal or resting metabolic rate values rather than actual metabolic rates of animals during the water flux measurements. The WEIs of free-living animals can now be readily determined by means of the doubly labeled water method, which simultaneously measures water flux and energy metabolism (Lifson and McClintock 1966, Nagy 1980). We assembled all available data on doubly labeled water measurements in free-living animals, and calculated WEI values (Table A3). We selected for subsequent analyses only those WEI values for animals that were maintaining constant body masses over the measurement period, in an attempt to eliminate variation due to nonsteady-state conditions. A constant body mass does not necessarily mean that an animal has a constant body composition (especially body water volume and fat stores), but it is often the only indication available in field studies. Water flux (and consequently WEI) may vary widely in nonsteady-state circumstances, and can be quite low (e.g., when an animal is starving and not drinking), or very high (e.g., when moist food or much water is being consumed). Even when an animal is maintaining water balance, it may be consuming much more water than it needs while excreting the excess as dilute urine. Thus, WEI values for animals maintaining constant body masses do not necessarily reflect their minimum water requirements.

## ENDOTHERMS AND ECTOTHERMS

WEI values for free-living animals maintaining constant body masses do not differ significantly between eutherians, marsupials, or birds (all $P > 0.05$ via Mann-Whitney $U$ test for data sets with unequal variances). Similarly, the mean of WEI values for reptiles does not differ from that for arthropods (Fig. 18), but the combined WEI values for ectotherms have a significantly ($P < 0.05$) different mean (0.270) than that for the combined WEI values for endotherms (0.200). Because ectotherms have higher, rather than lower, WEI values than do endotherms (Fig. 18), the hypothesis that ectotherms conserve water more effectively than do endotherms is not supported by these results.

We feel that water flux values and WEI values are most useful in understanding the water relations of animals when these two ways of evaluating water economy are combined. The relatively high absolute water fluxes of endotherms mean that they must obtain much more water each day from their surroundings than must ectotherms.

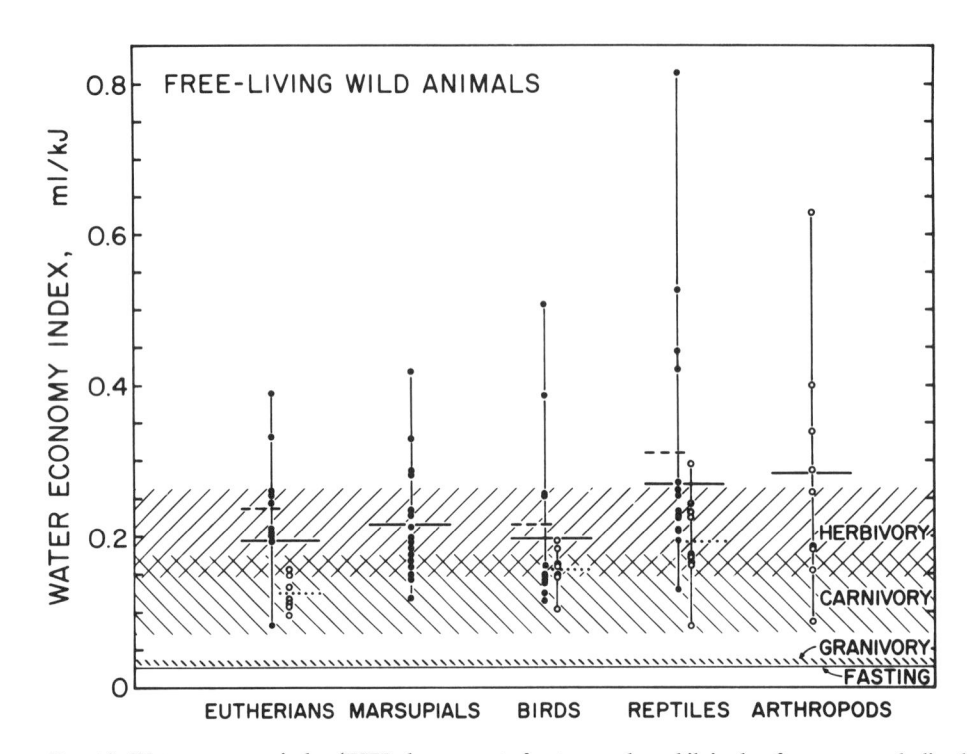

FIG. 18. Water economy index (WEI, the amount of water used per kilojoule of energy metabolized by animals) for air-breathing animals maintaining constant body masses in the field. Solid horizontal lines show means for each taxon; dotted and dashed horizontal lines indicate means for desert (o) and non-desert animals (●), respectively, within taxa. Cross-hatched areas indicate expected values of WEI for nondrinking animals maintaining water and energy balance on a diet of leaves (62 to 72% water content, to represent the food of a herbivore), animal matter (*Tenebrio* larvae to whole vertebrates, to represent the food of a carnivore), or mature seeds (0 to 10% water, for a granivore), and "fasting" line indicates the WEI for fat catabolism.

The more "urgent" water needs of endotherms might be expected to result in greater selection pressures favoring physiological, behavioral, and morphological adaptations that reduce water losses and thus decrease water requirements. The significantly lower WEI values in endotherms, compared with ectotherms, indicates that endotherms in general are more effective at conserving water, relative to their "rates of living" (metabolic rates). On the other hand, ectotherms, by virtue of their lower daily water needs, can afford to hide and wait out periods of adverse environmental conditions.

## DESERT AND NONDESERT ANIMALS

Within free-living eutherians, birds, and reptiles, WEI values tend to be lower in desert-dwelling species than in nondesert species (Fig. 18), although mean WEI values differ significantly only within eutherians ($P < 0.02$, Mann-Whitney $U$ test). Our hypothesis that desert vertebrates conserve water more effectively than their nondesert relatives is supported by the results for eutherians, and the results for birds and reptiles lend suggestive but not statistical support for this hypothesis. This indicates that desert animals employ, in their natural habitats, at least some of the physiological, behavioral, and morphological mechanisms for water conservation that have been abundantly documented in captive animals in the laboratory (Schmidt-Nielsen 1964, Maloiy 1972 and 1979, Gilles 1979, Louw and Seely 1982).

Among the data sets for desert animals (Fig. 18), eutherians and birds do not differ, nor do reptiles and arthropods, but the desert endotherms combined have a significantly lower mean WEI than do the desert ectotherms combined. The argument (above) that the higher daily water demands of endotherms should favor selection of lower WEIs may apply more strongly in the case of desert animals. The remarkable variation in WEI among free-ranging desert arthropods (Fig. 18) may reflect their behavioral adaptations for obtaining water from the diverse sources afforded by the numerous microhabitats available to very small desert animals, as well as season effects, among other factors. Desert arthropods apparently can achieve water balance by finding and consuming much water, as well as by using their unusual abilities to conserve water when it is scarce (Edney 1977).

## BODY MASS EFFECT

We have tested the WEI values for possible effects of body mass (Fig. 19) by using a variety of statistical procedures, including regression analyses of log-transformed and untransformed data. No convincing correlations have emerged from these efforts. No animals larger than 10 kg have WEI values lower than 0.175, and WEI values lower than 0.150 occur only in small animals ($<500$ g in birds and $<100$ g in mammals). However, some small animals have relatively high WEIs. Small size may be a prerequisite for having a low WEI, but small size does not mandate water conservation.

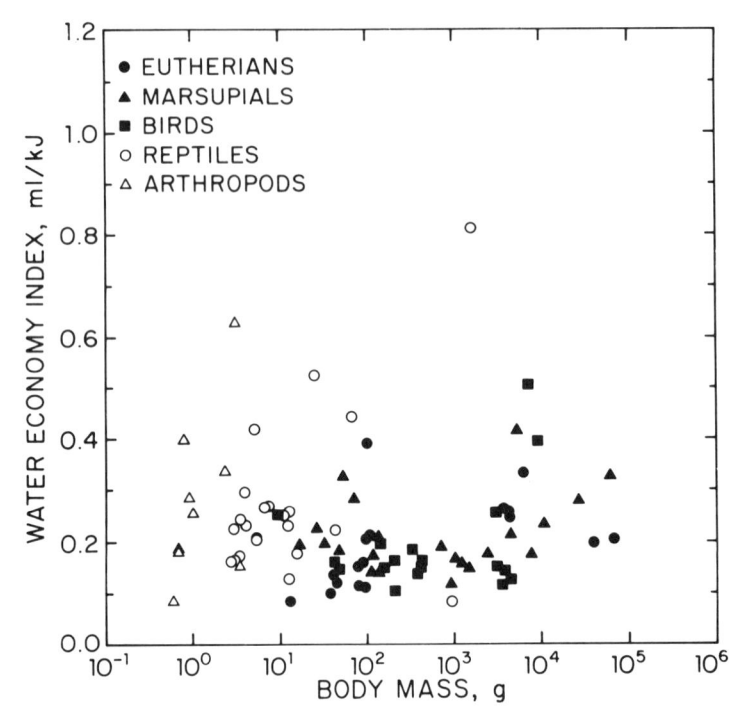

FIG. 19. Relationship of water economy index (WEI) in free-living, steady-state animals to body mass (on a logarithmic axis).

## WEI INTERPRETATION

A theoretical framework, useful for evaluating WEI values of terrestrial animals, can be derived by making some assumptions about drinking and diet composition. If an animal does not drink water, then its water intake consists essentially of dietary water, both preformed water in the food and oxidation (metabolic) water produced during catabolism of energy-containing chemicals in the food. Thus, the WEI value for a nondrinking terrestrial animal maintaining steady-state water and energy balance is the water yield of the diet divided by the energy yield of the diet. We used water and energy yield information (Nagy 1983b) to calculate WEIs for some typical diets: vegetation having water contents of 62% of fresh mass (representative of desert plants) and 72% (tropical plants), animal matter (lipid-rich and highly digestible *Tenebrio* larvae, and lean whole vertebrates), and seeds having preformed water contents of 0% of fresh mass and 10% (typical air-dried seeds). We also calculated WEI for an animal that is not drinking or eating and is metabolizing only body fat. The results of these calculations are shown as hatched areas in Figure 18.

WEI values for many nondesert animals exceed the highest value expected for nondrinking animals eating typical diets, indicating that they either were drinking water or were selecting especially succulent diets, or both (Fig. 18). Most of the desert vertebrates had WEIs within the range expected for herbivores and carnivores in the absence of drinking. Interestingly, no WEIs for free-living, steady-state animals were as low as expected for nondrinking seed-eaters. Kangaroo rats (*Dipodomys merriami*) are quite capable of maintaining water balance on a diet of air-dried seeds alone in captivity (Schmidt-Nielsen 1964). They do this not by producing large amounts of "metabolic" water via oxidation of foodstuffs (metabolic water production depends on metabolic rate, and kangaroo rats do not have notably high metabolic rates), but by conserving water so effectively that they require very little to maintain balance. This species has been studied in the field with doubly labeled water (Mullen 1971), but its WEIs are not shown in Figure 18 because no information about body mass maintenance or steady-state water balance was reported. Its WEI values (0.049 to 0.113) are higher than that expected for air-dried seeds (0.037), suggesting that these animals are obtaining more water in the field (probably by ingesting more succulent food) than expected from laboratory studies.

To date, no birds that are known to survive on dry seeds alone in captivity (Bartholomew 1972) have been studied in the field with doubly labeled water, and no desert marsupials have been studied. Animals that breathe water are not suitable for doubly labeled water studies because diffusional exchange of isotopic water across respiratory surfaces is rapid, and water flux rates are so large relative to $CO_2$ production rate that the difference between washout rates of the two isotopes (which measures $CO_2$ production, hence metabolic rate) is too small to resolve accurately. We used published values for resting metabolism (Prosser 1973) to estimate that WEI values in fish, water-breathing arthropods, and mollusks range between about 400 and 900 ml/kJ, more than 3 orders of magnitude higher than air-breathing animals. Most of this difference is due to diffusional exchange of water across respiratory surfaces and integument of aquatic animals.

# CONCLUSIONS

1. Rates of water flux, measured with isotopically labeled water, scale significantly with body mass ($r^2$ values for $\log_{10}$-$\log_{10}$ regressions range from 0.64 to 0.96) among eutherian mammals, marsupial mammals, birds, reptiles, fishes, water-breathing arthropods, air-breathing arthropods, and mollusks.
2. Regressions for free-living animals differ significantly in either slope or intercept (ANCOVA) from regressions for captive and domestic animals in all cases where sample sizes are adequate (eutherian mammals, marsupial mammals, birds, and reptiles). Ecological conclusions based on water flux studies of captive animals should be applied with caution. By manipulating experimental conditions, captive animals can be made to have water fluxes spanning an impressively large range (up to 70-fold in one species of eutherian). Variation in water flux rates of free-living animals can also be large, and is greatest among air-breathing arthropods, which apparently possess a rich diversity of water balance adaptations.
3. Water-breathing animals (annelid worms, mollusks, some arthropods, and fish) have the highest water fluxes; air-breathing endotherms (eutherians, marsupials, and birds) are intermediate; and air-breathing ectotherms (reptiles and some arthropods) have the lowest water fluxes. Significant allometric differences exist between taxa within these general categories, in captivity as well as in the field.
4. Differences exist in water flux scaling among captive animals within major taxonomic groups in accord with differences in taxonomic class (marine osteichthyes vs. marine chondrichthyes) and habitat (marine vs. freshwater fishes, marine vs. freshwater water-breathing arthropods, aquatic vs. mesic vs. desert air-breathing arthropods).

Conclusions

5. Among free-living animals, differences exist within major taxonomic groups in accord with differences in taxonomic order (passerine birds vs. all other birds); habitat (desert vs. nondesert eutherians, desert vs. nondesert birds, seabirds vs. other birds, desert vs. nondesert reptiles); and diet (herbivorous vs. other eutherians, herbivorous vs. carnivorous marsupials, and carnivorous vs. other birds).
6. Equations are given for predicting water flux rates for captive and free-living animals.
7. The "water economy index" (WEI), the amount of water used per unit of energy metabolized (ml/kJ, as measured using doubly labeled water in free-living animals maintaining constant body masses in the field), is similar among terrestrial endotherms and ectotherms, indicating that the large differences in absolute water fluxes between these groups are due primarily to differences in energy metabolism. However, within taxa, the lower water fluxes in desert species are not completely accounted for by their lower metabolic rates: desert species tend to have low WEI values compared to their nondesert relatives, indicating their use of water-conserving physiological, morphological, and behavioral adaptations in their natural habitats.
8. Comparison of actual WEI values determined in the field with those expected for nondrinking animals eating various diets indicates that many species are able to obtain additional water, but no species yet studied has accomplished the feat of surviving in the field while eating a diet of dry seeds with no other source of water.

# Appendices

TABLE A1. Summary of water flux rates, measured using tritiated or deuterated water, in captivity (including domestic animals) and in the field.

| GROUP<br>Species (diet, habitat)[1] | Body mass, g | Water flux rate ml/day | % of predicted[2] | Conditions | Reference |
|---|---|---|---|---|---|
| **EUTHERIAN MAMMALS** | | | | | |
| *In captivity* | | | | | |
| White mouse (O,M) | | | | | |
| *Mus musculus* | 21.4 | 7.7 | 270 | | Richmond et al. 1962 |
| | 32.2 | 5.9 | 140 | | Rouffignac and Morel 1966 |
| | 20* | 8.2 | 300 | | Thompson 1952 |
| | 56* | 11.2 | 160 | females | Ueno and Kawamura 1975 |
| | 26.0 | 4.2 | 120 | control strain | Chemama et al. 1972 |
| | 39.2 | 4.4 | 86 | predisposed to cancer | |
| | 26.0 | 5.2 | 150 | predisposed to cancer | |
| | 64 | 7.33 | 90 | genetically obese mice | McClintock and Lifson 1957 |
| | 26 | 8.90 | 260 | milk diet, diarrheic | McClintock and Lifson 1958a |
| | 25 | 3.1 | 93 | | Lifson et al. 1955 |
| | 20* | 4.5 | 170 | high humidity | Foy 1964 |
| | 20* | 6.7 | 250 | | Brues et al. 1952 |
| | 20* | 4.3 | 160 | | Arai et al. 1975 |

42

Table A1, continued (Eutherian mammals, captive)

| | | | | | |
|---|---|---|---|---|---|
| White mouse (cont.) | 21.0 | 4.7 | 170 | water ad lib. | Haines et al. 1973a |
| | 19.6 | 2.7 | 102 | water 1/2 ad lib. | |
| | 15.2 | 2.0 | 96 | water 1/4 ad lib. | |
| | 14.8 | 1.7 | 84 | water 1/8 ad lib. | |
| | 6.02 | 0.66 | 76 | 10 day old, nursing | Rath and Thenen 1979 |
| | 7.72 | 0.76 | 69 | 15 day old, nursing | |
| | 16.7 | 4.47 | 196 | water ad lib. | Sicard et al. 1985 |
| | 15.7 | 3.44 | 160 | water restricted | |
| | 14.9 | 3.36 | 160 | no water | |
| | 11.0 | 3.82 | 250 | water ad lib., rehydrating | |
| Wild mouse (O,X) | | | | | |
| Mus spretus | 14.5 | 3.32 | 170 | water ad lib. | Sicard et al. 1985 |
| | 13.7 | 2.56 | 140 | water restricted | |
| | 12.4 | 2.00 | 120 | no water | |
| | 14.0 | 3.46 | 180 | water ad lib., rehydrating | |
| Shrew (C,M) | | | | | |
| Blarina brevicauda | 22.9 | 16.2 | 530 | 20°C, water ad lib. | Deavers and Hudson 1977 |
| | 18.4 | 9.9 | 400 | 20°C, maintenance water | |
| | 23.4 | 25.4 | 810 | 5°C, water ad lib. | |
| | 18.3 | 18.7 | 750 | 5°C, maintenance water | |
| Rock mouse (O,X) | | | | | |
| Petromyscus collinus | 19 | 1.4 | 54 | dry food only | Withers et al. 1980 |
| | 19 | 2.9 | 110 | dry and wet food | |

1 Diets are: herbivore (H), carnivore (C), omnivore (O), granivore (G), nectarivore (N); habitats are: seawater (SW), fresh water (FW), xeric (X = arid, semi-arid terrestrial), mesic (M = temperate, subtropical) and hygric (H = swamp, rainforest). Habitat only is given for amphibians, fishes, and invertebrates.

2 Percent of predicted = 100 (actual ml/day)/(predicted ml/day), where predicted ml/day was calculated using the "in captivity" or "in field" allometric equations (Tables 1 and A2) for the major taxonomic groups.

* Body mass estimated.

Table A1, continued (Eutherian mammals, captive)

| GROUP Species (diet,habitat)[1] | Body mass, g | Water flux rate ml/day | % of predicted[2] | Conditions | Reference |
|---|---|---|---|---|---|
| White-footed mouse (O,M) *Peromyscus leucopus* | 22.2 | 6.2 | 210 | 20°C, water ad lib. | Deavers and Hudson 1977 |
| | 18.3 | 3.5 | 140 | 20°C, maintenance water | |
| | 23.9 | 8.6 | 270 | 5°C, water ad lib. | |
| | 19.5 | 5.6 | 210 | 5°C, maintenance water | |
| Deer mouse (O,X) *Peromyscus maniculatus* | 25* | 3.15 | 94 | | Grubbs 1980 |
| | 18.7 | 3.7 | 150 | | Holleman and Dieterich 1973 |
| Pinyon mouse (G,M) *Peromyscus truei* | 20.2 | 6.30 | 230 | water ad lib. | Bradford 1974 |
| | 20.2 | 5.38 | 190 | maintenance water | |
| | 20.2 | 1.84 | 67 | minimum water | |
| Pocket mouse (G,X) *Perognathus penicillatus* | 18* | 1.90 | 78 | 10°C | Grubbs 1980 |
| | 18* | 1.49 | 61 | 20°C | |
| | 18* | 0.95 | 39 | 23°C | |
| | 13 | 1.20 | 67 | 23°C | |
| | 18 | 0.71 | 29 | 30°C | |
| | 18 | 2.05 | 84 | 10°C, no water | MacMillen and Hinds 1983 |
| | 18 | 1.59 | 65 | 20°C, no water | |
| | 18 | 0.79 | 32 | 30°C, no water | |
| Pocket mouse (G,X) *Perognathus fallax* | 21 | 2.57 | 91 | 10°C, no water | MacMillen and Hinds 1983 |
| | 21 | 1.60 | 56 | 23°C, no water | |
| | 21 | 0.96 | 34 | 30°C, no water | |

Table A1, continued (Eutherian mammals, captive)

| | | | | | |
|---|---|---|---|---|---|
| Pocket mouse (G,X) | | | | | |
| Perognathus formosus | 18.6 | 0.6 | 24 | dry seeds, no water | Mullen 1970 |
| Vesper mouse (H,M) | | | | | |
| Calomys ducilla | 26.6 | 3.1 | 88 | | Holleman and Dieterich 1973 |
| Hamster (O,X) | | | | | |
| Phodopus sungorus | 28.5 | 5.9 | 160 | rehydrated after water restriction | Schierwater and Klingel 1985 |
| | 30.3 | 3.7 | 92 | normally hydrated | |
| Gerbil (O,X) | | | | | |
| Gerbillus gerbillus | 34.2 | 4.5 | 100 | | Rouffignac and Morel 1966 |
| Hopping mouse (O,X) | | | | | |
| Notomys alexis | 30.0 | 5.4 | 140 | water ad lib. | Hewitt et al. 1981 |
| | 24.4 | 1.2 | 37 | dehydrated, water ad lib. | |
| | 20.6 | 1.4 | 50 | no water | |
| | 29.0 | 5.8 | 150 | rehydrated, no water | |
| | 49.6 | 2.6 | 41 | high protein diet, no water | |
| | 35 | 6.65 | 140 | | Macfarlane and Howard 1972 |
| | 33 | 7.38 | 170 | | Haines et al. 1973b |
| | 34.7 | 6.7 | 150 | water ad lib. | Haines et al. 1974 |
| | 24.6 | 1.2 | 36 | water restricted | |
| Hopping mouse (O,X) | | | | | |
| Notomys cervinus | 41.2 | 7.1 | 130 | water ad lib. | Haines et al. 1974 |
| | 30.7 | 1.6 | 39 | water restricted | |
| | 41 | 6.11 | 110 | | Macfarlane and Howard 1972 |

Table A1, continued (Eutherian mammals, captive)

| GROUP Species (diet, habitat)[1] | Body mass, g | Water flux rate ml/day | % of predicted[2] | Conditions | Reference |
|---|---|---|---|---|---|
| Lemming (H,M) | | | | | |
| Dicrostonyx groenlandicus | 44.7 | 10.4 | 180 | | Holleman and Dieterich 1973 |
| Rock mouse (O,X) | | | | | |
| Aethomys namaquensis | 46 | 1.7 | 29 | dry food only | Withers et al. 1980 |
| | 46 | 4.8 | 81 | dry and wet food | |
| Vole (H,H) | | | | | |
| Microtus pennsylvanicus | 21.8 | 8.1 | 280 | | Holleman and Dieterich 1973 |
| Vole (H,H) | | | | | |
| Microtus oeconomus | 31.0 | 12.0 | 290 | | Holleman and Dieterich 1973 |
| Vole (H,M) | | | | | |
| Microtus abbreviatus | 50.0 | 14.3 | 220 | | Holleman and Dieterich 1973 |
| Vole (H,M) | | | | | |
| Microtus californicus | 40.1 | 10.7 | 200 | water ad lib. | Church 1966 |
| | 35.7 | 6.9 | 150 | water 2/3 ad lib. | |
| | 30.1 | 5.3 | 130 | water 1/3 ad lib. | |
| | 27.9 | 3.5 | 94 | water less than 1/3 ad lib. | |
| Vole (O,M) | | | | | |
| Clethrionomys gapperi | 25.7 | 13.8 | 400 | 20°C, water ad lib. | Deavers and Hudson 1977 |
| | 19.6 | 7.8 | 290 | 20°C, maintenance water | |
| | 22.2 | 21.2 | 710 | 5°C, water ad lib. | |
| | 17.7 | 13.2 | 550 | 5°C, maintenance water | |
| Vole (H,M) | | | | | |
| Microtus ochrogaster | 46.2 | 11.7 | 200 | water ad lib. | Dupre 1983 |
| | 35.4 | 7.6 | 160 | maintenance water | |

Table A1, continued (Eutherian mammals, captive)

| | | | | |
|---|---|---|---|---|
| Vole (O,M) | | | | |
| Clethrionomys rutilus | 22.6 | 6.4 | 210 | winter, 18°C | Holleman et al. 1982 |
| | 23.6 | 6.1 | 190 | spring, 18°C | |
| | 27.1 | 10.9 | 300 | summer, 18°C | |
| | 29.0 | 8.2 | 210 | autumn, 18°C | |
| Vole (H,H and M) | | | | | |
| Arvicola terrestris | 40.6 | 38.1 | 720 | juveniles | Grenot et al. 1984 |
| | 57.4 | 46.2 | 630 | subadults | |
| Spiny mouse (O,X) | | | | | |
| Acomys cahirinus | 49.0 | 4.8 | 76 | | Holleman and Dieterich 1973 |
| | 44.3 | 6.6 | 110 | water ad lib. | Daily and Haines 1981 |
| | 30.5 | 1.7 | 42 | water 1/4 ad lib. | |
| | 24.9 | 1.3 | 39 | water 1/8 ad lib. | |
| | 50* | 8.2 | 130 | | Pietz 1974 |
| Eastern mouse (O,X) | | | | | |
| Pseudomys australis | 50 | 6.95 | 108 | | Macfarlane and Howard 1972 |
| | 49.9 | 6.5 | 101 | water ad lib. | Haines et al. 1974 |
| | 32.3 | 2.0 | 47 | water restricted | |
| Desert mouse (O,X) | | | | | |
| Pseudomys desertor | 43.9 | 8.9 | 160 | water ad lib. | Haines et al. 1974 |
| | 31.3 | 2.2 | 53 | water restricted | |
| Spiny pocket mouse (O,X) | | | | | |
| Liomys irroratus | 51.8 | 6.42 | 96 | | Hudson and Rummel 1966 |
| Liomys salvani | 54.5 | 6.59 | 94 | | |
| Liomys sp. | 53* | 4.16 | 61 | no water, burrowable soil | |
| | 53* | 2.78 | 41 | no water, open cage | |

47

Table A1, continued (Eutherian mammals, captive)

| GROUP Species (diet,habitat)[1] | Body mass, g | Water flux rate ml/day | Water flux rate % of predicted[2] | Conditions | Reference |
|---|---|---|---|---|---|
| Chipmunk (O,M) Eutamias palmeri | 69.4 | 21.5 | 240 | food ad lib., no water | Yousef et al. 1974 |
| Chipmunk (O,M) Tamias striatus | 106 | 11.9 | 91 | | Little and Lifson 1975 |
| Wood rat (H,X) Neotoma lepida | 98.9 | 15.2 | 120 | food ad lib., no water | Yousef et al. 1974 |
| Kangaroo rat (G,X) Dipodomys merriami | 33.9 | 1.2 | 27 | food ad lib., no water | Yousef et al. 1974 |
| | 35.5* | 2.14 | 46 | | Hatch and Mazrimas 1972 |
| | 37 37 37 | 2.80 1.89 1.09 | 58 39 23 | 10 C, no water 23 C, no water 30 C, no water | MacMillen and Hinds 1983 |
| | 40* | 1.98 | 38 | | Grubbs 1980 |
| Kangaroo rat (O,X) Dipodomys microps | 54.2 | 3.3 | 48 | food ad lib., no water | Yousef et al. 1974 |
| Kangaroo rat (G,X) Dipodomys ordii | 51* | 1.66 | 25 | | Martin and Koranda 1972 |
| Kangaroo rat (G,X) Dipodomys deserti | 93 | 3.3 | 29 | dry seeds, no water | Richmond et al. 1962 |
| | 97.2 99.8 | 4.32 2.92 | 36 24 | 20-30% rh 10-20% rh | Richmond et al. 1960 |
| | 101 | 3.2 | 26 | food ad lib., no water | Yousef et al. 1974 |

Table A1, continued (Eutherian mammals, captive)

| Species | | | | Reference |
|---|---|---|---|---|
| Pocket gopher (H,M) *Thomomys bottae* | 127 | 11 | 71 | no water | Gettinger 1983 |
| Sand rat (O,X) *Meriones shawii* | 159<br>121<br>96 | 13.2<br>9.2<br>2.6 | 69<br>62<br>22 | | Rouffignac and Morel 1966 |
| | 201.3<br>143.3 | 50.1<br>13.0 | 208<br>75 | no water<br>mixed diet, no water<br>seeds, no water | Bradshaw et al. 1976b |
| Sand rat (O,X) *Meriones unguiculatus* | 80.3 | 6.7 | 67 | | Holleman and Dieterich 1973 |
| Sand rat (O,X) *Meriones crassus* | 127.3 | 16.0 | 103 | seeds, no water | Bradshaw et al. 1976b |
| Ground squirrel (O,X) *Spermophilus tereticaudus* | 117 | 14.1 | 98 | food ad lib., no water | Yousef et al. 1974 |
| Ground squirrel (O,M) *Spermophilus lateralis* | 61.8 | 27.9 | 350 | food ad lib., no water | Yousef et al. 1974 |
| Ground squirrel (O,M) *Spermophilus tridecemlineatus* | 200*<br>200* | 14.5<br>0.2 | 61<br>1 | cold-acclimated, normothermic hibernating | Deavers and Musacchia 1976 |
| Ground squirrel (O,X) *Ammospermophilus leucurus* | 88.1<br>78.0<br>73.0<br>80.9<br>70.4<br>82.0* | 14.2<br>13.5<br>3.6<br>11.6<br>4.7<br>5.5 | 130<br>140<br>39<br>110<br>53<br>54 | no water, insect diet<br>no water, plant diet<br>no water, seed diet, carrot supplement<br>water ad lib, lab chow<br>water restricted, lab chow<br>outdoor enclosure, autumn, water restricted | Karasov 1983a |
| | 86.3 | 12.5 | 120 | food ad lib., no water | Yousef et al. 1974 |

Table A1, continued (Eutherian mammals, captive)

| GROUP Species (diet,habitat)[1] | Body mass, g | Water flux rate ml/day | % of predicted[2] | Conditions | Reference |
|---|---|---|---|---|---|
| Cotton rat (O,M) Sigmodon hispidus | 181 167 | 36.1 19.1 | 170 95 | water ad lib. maintenance water | Dupre 1983 |
| White rat (O,M) Rattus norvegicus | 298 | 34.9 | 100 | | Richmond et al. 1962 |
| | 221 169 | 26.3 19.8 | 100 97 | | Rouffignac and Morel 1966 |
| | 371 | 58.2 | 136 | | Holleman and Dieterich 1973 |
| | 220 | 29.0 | 110 | | McClintock and Lifson 1958b |
| | 300 | 39.3 | 110 | | Boxer and Stetten 1944 |
| | 228 | 26.4 | 98 | | Lee and Lifson 1960 |
| | 200 | 11.6 | 49 | no food, no water | Lifson and Lee 1961 |
| | 175 | 25.7 | 120 | females | Thompson 1953 |
| | 450* | 45.6 | 89 | males | Wheeler et al. 1972 |
| | 260 | 21.9 | 72 | | Gleason and Friedman 1970 |
| | 170 | 31.9 | 160 | | Bogdanov et al. 1958 |
| | 135 | 26.5 | 160 | | Lambert and Clifton 1967 |

Table A1, continued (Eutherian mammals, captive)

| | | | | |
|---|---|---|---|---|
| White rat (cont.) | 200* | 28.7 | 120 | high humidity | Foy 1964 |
| | 200* | 26.4 | 110 | tropics, shade, distilled water | |
| | 200* | 33.8 | 140 | tropics, shade, saline water | |
| | 200* | 25.8 | 108 | moderate humidity, distilled water | |
| | 200* | 33.0 | 140 | moderate humidity, saline water | |
| | 200* | 29.4 | 120 | low humidity, low temp., dist. water | |
| | 200* | 37.3 | 160 | low humidity, low temp., saline water | |
| | 200* | 19.0 | 80 | | Peitz 1974 |
| | 14.5 | 1.58 | 80 | 7 day old, nursing | Romero et al. 1975 |
| | 19.4 | 3.96 | 150 | 14 day old, nursing | |
| | 390 | 51.7 | 120 | males | Takeda 1982 |
| Stick-nest rat (H,X) Leporillus conditor | 327 | 43.7 | 110 | water ad lib. | Haines et al. 1974 |
| Chinchilla (H,M) Chinchilla laniger | 412 | 26.7 | 56 | | Holleman and Dieterich 1973 |
| Guinea pig (O,M) Cavia porcellus | 600* | 64.3 | 95 | high humidity | Foy 1964 |
| | 680 | 56.7 | 75 | | Holleman and Dieterich 1973 |
| Muskrat (H,H) Ondatra zibethica | 753 | 152 | 180 | | Holleman and Dieterich 1973 |
| Prairie dog (H,M) Cynomys sp. | 1000* | 46.5 | 43 | | Peitz 1974 |
| Kit fox (O,X) Vulpes macrotis | 1850 | 205 | 105 | no food or water | Golightly and Ohmart 1984 |
| | 1850 | 31 | 16 | | |

51

Table A1, continued (Eutherian mammals, captive)

| GROUP Species (diet,habitat)[1] | Body mass, g | Water flux rate ml/day | % of predicted[2] | Conditions | Reference |
|---|---|---|---|---|---|
| Jackrabbit (H,X) |  |  |  |  |  |
| *Lepus californicus* | 1800 | 234 | 120 | spring, water ad lib. | Nagy et al. 1976 |
|  | 1800 | 326 | 170 | summer, water ad lib. |  |
|  | 1800 | 149 | 78 | winter, water ad lib. |  |
|  | 2270 | 192 | 81 | water ad lib. | Reese and Haines 1978 |
|  | 2150 | 90 | 40 | water restricted |  |
| Rabbit (H,M) |  |  |  |  |  |
| *Oryctolagus cuniculus* | 3160 | 328 | 101 |  | Richmond et al. 1962 |
|  | 1800 | 123 | 64 |  | Green and Dunsmore 1978 |
|  | 2000* | 323 | 150 | high humidity | Foy 1964 |
| Rock hyrax (H,X) |  |  |  |  |  |
| *Procavia habessinica* | 2400 | 126 | 50 | 20°C, food, water ad lib. | Rubsamen et al. 1979 |
|  | 2400 | 90 | 36 | 27°C, food, water ad lib. | Rubsamen and Engelhardt 1982 |
|  | 2400 | 113 | 45 | 30°C, food, water ad lib. |  |
|  | 2400 | 119 | 48 | 20°C/30°C, food, water ad lib. |  |
|  | 2400 | 255 | 102 | 20°C, food (dry, wet) ad lib., no water |  |
|  | 2400 | 78 | 31 | 27°C, food ad lib., no water |  |
|  | 2400 | 101 | 40 | 35°C, food ad lib., no water |  |
|  | 2400 | 53 | 21 | 20°C, dry food only |  |
|  | 2400 | 45 | 18 | 27°C, dry food only |  |
|  | 2400 | 60 | 24 | 35°C, dry food only |  |
|  | 2400 | 50 | 20 | 20°C/30°C, dry food only |  |
| Baboon (O,M) |  |  |  |  |  |
| *Papio cynocephalus* | 2500 | 588 | 226 | infant 4 months old | Buss and Voss 1971 |

Table A1, continued (Eutherian mammals, captive)

| | | | | |
|---|---|---|---|---|
| Cat (C,M) | | | | |
| Felis domesticus | 3000* | 174 | 56 | Wheeler et al. 1972 |
| | 3030 | 182 | 58 | dry food, water ad lib. |
| | 2810 | 152 | 52 | moist food, water ad lib. | Seefeldt and Chapman 1979 |
| Macaque (O,M) | | | | |
| Macaca fascicularis | 3200 | 195 | 59 | Kamis and Latif 1981 |
| | 3040 | 187 | 60 | Azar and Shaw 1975 |
| Gibbon (O,H) | | | | |
| Hylobates lar | 3850 | 678 | 170 | Kamis and Latif 1981 |
| Howler monkey (H,M) | | | | |
| Alouatta palliata | 5600 | 706 | 130 | natural diet | Nagy and Milton 1979b |
| Rhesus monkey (O,M) | | | | |
| Macaca mulatta | 5700 | 536 | 94 | males, 24°C |
| | 6200 | 360 | 59 | males, 6°C | Oddershede and Elizondo 1980 |
| | 4870 | 597 | 120 | | Azar and Shaw 1975 |
| Coyote (O,X) | | | | |
| Canis latrans | 10000 | 1250 | 130 | summer |
| | 10000 | 550 | 57 | winter |
| | 10000 | 88 | 9 | no food or water | Golightly and Ohmart 1984 |
| Dog (C,M) | | | | |
| Canis familiaris | 10600 | 942 | 92 | | Richmond et al. 1962 |
| | 14300 | 572 | 42 | | Gaebler and Choitz 1964 |
| Peccary (O,X) | | | | |
| Dicotyles tajacu | 22400 | 1570 | 76 | |
| | 20100 | 550 | 29 | water restricted | Zervanos and Day 1977 |

Table A1, continued (Eutherian mammals, captive)

| GROUP Species (diet,habitat)[1] | Body mass, g | Water flux rate ml/day | % of predicted[2] | Conditions | Reference |
|---|---|---|---|---|---|
| Fur seal (C,SW) Callorhinus ursinus | 18000 18000 18000 18000 | 522 576 864 936 | 31 34 51 56 | 10oC, water restricted 5oC, water restricted fresh water ad lib. fresh water ad lib. | Ohata et al. 1975 |
| Pronghorn antelope (H,M and X) Antilocapra americana | 22300 19200 | 1940 2420 | 94 140 | young males, metabolic cage young females, metabolic cage | Wesley et al. 1970 |
| Sea otter (C,SW) Enhydra lutris | 24300 | 6540 | 290 | | Costa 1982 |
| Harbor seal (C,SW) Phoca vitulina | 28600 31800 27800 | 515 1140 1640 | 20 40 65 | starved 1/2 food ration full food ration | Depocas et al. 1971 |
| Harbor porpoise (C,SW) Phocoena phocoena | 30000 | 1280 | 47 | feeding in 33% SW | Andersen and Nielsen 1983 |
| Deer (H,M) Odocoileus hemionus | 33300 32000 | 1750 3360 | 58 120 | winter summer | Longhurst et al. 1970 |
| | 20000 36000 | 1480 7780 | 79 240 | indoors in small cage | Knox et al. 1969 |
| Sea lion (C,SW) Zalophus californianus | 36300 | 3550 | 108 | | Costa 1984 |

Table A1, continued (Eutherian mammals, captive)

| | | | | |
|---|---|---|---|---|
| Goat (H,M) | | | | |
| Capra hircus | 70000 | 3640 | 60 | | Macfarlane and Howard 1972 |
| | 40000 | 3840 | 107 | | |
| | 22000 | 1740 | 85 | | Macfarlane et al. 1972 |
| | 35900 | 3220 | 99 | spring pasture | Macfarlane et al. 1974 |
| | 38200 | 2360 | 69 | autumn pasture | |
| | 13300 | 1440 | 110 | wet season | Aggrey 1982 |
| | 13600 | 1270 | 98 | dry season | |
| | 27900 | 3220 | 130 | summer | Dawson et al. 1975 |
| | 28500 | 3450 | 130 | dry season, males | El Hadi and Hassan 1982 |
| | 27000 | 5820 | 230 | wet season, males | |
| | 27000 | 2050 | 83 | minimum flux rate | King 1979 |
| | 27000 | 5290 | 210 | maximum flux rate | |
| | 21300 | 4480 | 230 | metabolic chamber, lactating | Maltz and Shkolnik 1980 |
| | 18600 | 1610 | 93 | metabolic chamber, non-lactating | |
| | 40600 | 3850 | 106 | dry tropics | Macfarlane 1965 |
| | 25300 | 1970 | 85 | 18oC | Kamal et al. 1972 |
| | 24100 | 3570 | 160 | 32oC | |
| | 14000 | 1220 | 92 | winter | Ranjhan et al. 1982 |
| | 16000 | 2380 | 160 | summer | |
| | 38800 | 1180 | 34 | food, water ad lib. | Rubsamen and Engelhardt 1975 |
| | 38800 | 825 | 24 | food restricted | |
| | 38800 | 849 | 24 | water restricted | |

Table A1, continued (Eutherian mammals, captive)

| GROUP Species (diet,habitat)[1] | Body mass, g | Water flux rate ml/day | % of predicted[2] | Conditions | Reference |
|---|---|---|---|---|---|
| Sheep (H,M) | | | | | |
| Ovis aries | 47500 | 4800 | 110 | winter, no water, 20 sheep/ha | Lynch et al. 1972 |
| | 50000 | 5000 | 110 | spring, no water, 20 sheep/ha | |
| | 50500 | 3690 | 82 | summer, no water, 20 sheep/ha | |
| | 46900 | 4270 | 102 | winter, water, 20 sheep/ha | |
| | 49100 | 5060 | 120 | spring, water, 20 sheep/ha | |
| | 48200 | 3710 | 87 | summer, water, 20 sheep/ha | |
| | 44200 | 4240 | 108 | winter, no water, 50 sheep/ha | |
| | 42400 | 5680 | 150 | spring, no water, 50 sheep/ha | |
| | 43600 | 2220 | 57 | summer, no water, 50 sheep/ha | |
| | 44400 | 4350 | 110 | winter, water, 50 sheep/ha | |
| | 43200 | 6090 | 160 | spring, water, 50 sheep/ha | |
| | 45700 | 5480 | 130 | summer, water, 50 sheep/ha | |
| | 41000 | 8300 | 230 | spring, no water, 20 sheep/ha | Brown and Lynch 1972 |
| | 42000 | 4500 | 120 | summer, no water, 20 sheep/ha | |
| | 37000 | 4900 | 150 | spring, no water, 40 sheep/ha | |
| | 34000 | 2400 | 78 | summer, no water, 40 sheep/ha | |
| | 45000 | 10900 | 270 | spring, water, 20 sheep/ha | |
| | 46000 | 6800 | 170 | summer, water, 20 sheep/ha | |
| | 31000 | 8500 | 300 | spring, water, 40 sheep/ha | |
| | 29000 | 6200 | 230 | summer, water, 40 sheep/ha | |
| | 16600 | 2380 | 150 | 1 month old, nursing, large breed | Macfarlane et al. 1974 |
| | 23500 | 2880 | 130 | 2 month old, large breed | |
| | 13700 | 2720 | 210 | 1 month old, small breed | |
| | 18400 | 3420 | 200 | 2 month old, small breed | |
| | 11100 | 1770 | 170 | 1 month old, single lambs | |
| | 13800 | 2190 | 170 | 2 month old, single lambs | |
| | 9700 | 1950 | 210 | 1 month old, twin lambs | |
| | 12500 | 2710 | 230 | 2 month old, twin lambs | |

Table A1, continued (Eutherian mammals, captive)

| | | | | |
|---|---|---|---|---|
| Sheep (cont.) | 62000 | 10600 | 200 | spring pasture | Macfarlane et al. 1974 |
| | 61000 | 4980 | 93 | autumn pasture | |
| | 21800 | 2790 | 140 | wet season | Aggrey 1982 |
| | 21500 | 2090 | 105 | dry season | |
| | 38700 | 5610 | 160 | Negev desert, shorn | Benjamin et al. 1977 |
| | 40600 | 5620 | 150 | Negev desert, shorn | |
| | 36800 | 3500 | 106 | Negev desert, winter, no water | Benjamin et al. 1975 |
| | 36200 | 3350 | 103 | Negev desert, winter, no water | |
| | 7000* | 1240 | 180 | 2 day old, nursing | Coward et al. 1982 |
| | 28200 | 4890 | 190 | summer | Dawson et al. 1975 |
| | 40100 | 5260 | 150 | Negev desert, nonpregnant | Degen 1977 |
| | 48600 | 7270 | 170 | Negev desert, nonpregnant | |
| | 49500 | 7120 | 160 | Negev desert, pregnant | Degen 1977 |
| | 42600 | 7540 | 200 | Negev desert, lactating | |
| | 56000 | 6930 | 140 | Negev desert, pregnant | |
| | 44800 | 6850 | 170 | Negev desert, lactating | |
| | 9800 | 1610 | 170 | 1 week old, nursing | Dove and Freer 1979 |
| | 19700 | 1650 | 90 | 10 week old | |
| | 7500 | 1530 | 210 | 1 week old single lambs | |
| | 18000 | 2380 | 140 | 7 week old single lambs | |
| | 6500 | 1500 | 230 | 1 week old twins | |
| | 17000 | 2350 | 150 | 7 week old twins | |
| | 36600 | 4230 | 130 | dry season, males, desert | El Hadi and Hassan 1982 |
| | 36400 | 7130 | 220 | wet season, males, desert | |
| | 39400 | 4250 | 120 | dry season, males, riverine | |
| | 39800 | 8510 | 240 | wet season, males, riverine | |
| | 50300 | 2720 | 61 | | Faichney and Boston 1985 |
| | 40000* | 4030 | 110 | dry summer, water | Kamal 1982 |

Table A1, continued (Eutherian mammals, captive)

| GROUP Species (diet,habitat)[1] | Body mass, g | Water flux rate mL/day | Water flux rate % of predicted[2] | Conditions | Reference |
|---|---|---|---|---|---|
| Sheep (cont.) | 29000 | 1800 | 68 | minimum flux rate | King 1979 |
|  | 29000 | 4840 | 180 | maximum flux rate |  |
|  | 42800 | 4460 | 120 |  | Knox et al. 1970 |
|  | 60600 | 5610 | 106 | dry saltbush pasture | Macfarlane et al. 1967 |
|  | 52400 | 5820 | 130 | grass pasture |  |
|  | 48100 | 9430 | 220 | saltbush pasture |  |
|  | 39700 | 6870 | 190 | grass pasture |  |
|  | 39400 | 13800 | 390 | saltbush pasture |  |
|  | 49100 | 3100 | 71 | temperate, winter | Macfarlane 1965 |
|  | 34200 | 4620 | 150 | tropics, summer, dry pasture |  |
|  | 35000 | 3310 | 105 | tropics, green pasture |  |
|  | 32200 | 2620 | 90 | wool intact | Macfarlane et al. 1966b |
|  | 32900 | 5230 | 180 | shorn |  |
|  | 51700 | 3760 | 82 |  | Anand et al. 1966 |
|  | 43300 | 4860 | 130 | high yield wool variety | Macfarlane et al. 1966a |
|  | 43200 | 4730 | 120 | control |  |
|  | 52800 | 3220 | 69 | autumn, 15 sheep/ha | Morris et al. 1962 |
|  | 53000 | 3330 | 71 | winter, 15 sheep/ha |  |
|  | 49900 | 4520 | 102 | winter, 15 sheep/ha, shorn |  |
|  | 51300 | 2230 | 49 | autumn, 22 sheep/ha |  |
|  | 43000 | 1390 | 36 | winter, 22 sheep/ha |  |
|  | 39300 | 2080 | 59 | winter, 22 sheep/ha, shorn |  |

Table A1, continued (Eutherian mammals, captive)

| Sheep (cont.) | 50700 | 5350 | 120 | winter | Longhurst et al. 1970 |
|---|---|---|---|---|---|
| | 55100 | 7810 | 160 | summer | |
| | 42000 | 8690 | 230 | saline water | Jones et al. 1970 |
| | 38300 | 3450 | 100 | fresh water | |
| | 57000 | 4980 | 99 | 18°C | Kammal et al. 1972 |
| | 55200 | 10300 | 210 | 32°C | |
| | 58000* | 5070 | 99 | eating saltbush, no shade | Wilson 1974 |
| | 62000* | 3540 | 65 | eating grass, no shade | |
| | 61000* | 3200 | 60 | eating grass, with shade | |
| | 62700 | 3040 | 55 | control | Phillips et al. 1969 |
| | 57900 | 2400 | 47 | acute hypobaria (350 mm Hg) | |
| | 58400 | 5500 | 107 | chronic hypobaria | |
| | 41700 | 5300 | 140 | adults, wet tropics | Macfarlane et al. 1974 |
| | 34800 | 4110 | 130 | nonlactating ewes | |
| | 7900 | 1820 | 240 | lambs | |
| | 39500 | 2500 | 71 | winter | Ranjhan et al. 1982 |
| | 45100 | 5000 | 120 | summer | |
| | 42200 | 3900 | 103 | winter | |
| | 46100 | 6100 | 150 | summer | |
| | 43100 | 3000 | 78 | winter | |
| | 41600 | 4990 | 130 | summer | |
| | 44600 | 5030 | 130 | winter | |
| | 48700 | 7040 | 160 | summer | |
| | 45300 | 4540 | 110 | nonpregnant, nonlactating | Russel et al. 1982 |
| | 36700 | 3520 | 106 | pregnant | |
| | 41300 | 8070 | 220 | lactating | |
| | 38200 | 4310 | 120 | summer pasture | Wright 1982a |
| | 9610 | 1520 | 160 | lambs nursing | Wright 1982a |

59

Table A1, continued (Eutherian mammals, captive)

| GROUP Species (diet,habitat)[1] | Body mass, g | Water flux rate mL/day | % of predicted[2] | Conditions | Reference |
|---|---|---|---|---|---|
| Sheep (cont.) | 20400 | 980 | 52 | 10°C, food rationed | Degen and Young 1981 |
| | 38400 | 2470 | 72 | 20°C, food rationed | |
| | 59900 | 6750 | 130 | 30°C, food rationed | |
| | 39500 | 1800 | 51 | winter | Kalanidhi et al. 1980 |
| | 45100 | 3650 | 91 | summer | |
| | 42200 | 2770 | 73 | winter | |
| | 46200 | 4400 | 107 | summer | |
| | 43200 | 2220 | 58 | winter | |
| | 41700 | 3630 | 97 | summer | |
| Dolphin (C,SW) Delphinus delphis | 57000 | 4390 | 88 | fasted | Hui 1981 |
| Bighorn sheep (H,X) Ovis canadensis | 65000* | 4850 | 85 | spring | Yousef 1971 |
| | 65000* | 7010 | 120 | summer | |
| Human (O,M) Homo sapiens | 67300 | 2760 | 47 | U.S. diet | Richmond et al. 1962 |
| | 60000* | 2280 | 43 | hospitalized | Olsson 1970 |
| | 60000* | 3780 | 72 | females in high humidity | Ray and Burch 1959 |
| | 58200 | 2330 | 46 | female university students | Schloerb et al. 1950 |
| | 71800 | 3420 | 55 | male university students | |
| | 66500 | 2860 | 49 | | Snyder et al. 1968 |
| | 74600 | 3470 | 54 | males | Wylie et al. 1963 |

Table A1, continued (Eutherian mammals, captive)

| | | | | | |
|---|---|---|---|---|---|
| Human (cont.) | 5810 | 848 | 150 | infant, nursing | Butte et al. 1983 |
| | 8000* | 958 | 120 | 3 month old, nursing | Coward et al. 1979 |
| | 60000* | 2910 | 55 | | Hevesy and Hofer 1934 |
| | 73500 | 2880 | 45 | water: 2.7 liters/day | Pinson and Langham 1957 |
| | 73500 | 12200 | 190 | water: 12.8 liters/day | |
| | 64700 | 2420 | 43 | water ad lib. | |
| | 55000 | 3600 | 74 | humid tropics, Nigerians | Foy and Schnieden 1960 |
| | 63000 | 6300 | 110 | Bantu miners working in humid heat | Macfarlane et al. 1966c |
| | 70000* | 3420 | 56 | hospitalized | Fallot et al. 1957 |
| | 53100 | 2760 | 59 | winter, tropics, Chimbu | Macfarlane and Howard 1966a |
| | 67300 | 2760 | 47 | winter, tropics, caucasian | |
| | 70000* | 2890 | 47 | | Haley et al. 1951, 1953 |
| | 70000* | 3440 | 56 | | Chiswell and Dancer 1969 |
| | 70000* | 3530 | 58 | spring and summer, North America | Butler and Leroy 1965 |
| | 70000* | 2810 | 46 | autumn and winter, North America | |
| | 62500 | 2810 | 51 | pregnant females | Hutchinson et al. 1954 |
| | 61100 | 4050 | 76 | post-partum females | |
| | 70000 | 2000 | 33 | without beer | Moore et al. 1968 |
| | 70000 | 4850 | 80 | with beer | |
| | 57500 | 2400 | 47 | | Haines et al. 1974 |
| | 70000* | 4940 | 81 | | Roberts et al. 1958 |
| | 77100 | 3330 | 50 | | Schoeller and van Santen 1982 |

Table A1, continued (Eutherian mammals, captive)

| GROUP Species (diet,habitat)[1] | Body mass, g | Water flux rate mL/day | % of predicted[2] | Conditions | Reference |
|---|---|---|---|---|---|
| Human (cont.) | 72900 | 3910 | 62 | metabolism chamber | Schoeller and Webb 1984 |
|  | 63000 | 3400 | 62 | Europeans, Australian desert | Macfarlane 1969 |
|  | 47000 | 5400 | 130 | Aboriginal males, Australian desert |  |
|  | 46800 | 5800 | 140 | Aboriginal females, lactating |  |
| Hartebeest (H,M and X) Alcelaphus buselaphus | 88000 | 4580 | 61 |  | Macfarlane and Howard 1972 |
|  | 87600 | 4560 | 61 |  | Maloiy and Hopcraft 1971 |
| Reindeer (H,M) Rangifer tarandus | 100000 | 12800 | 150 |  | Macfarlane and Howard 1972 |
|  | 80200 | 2330 | 34 | winter | Cameron and Luick 1972 |
|  | 86000 | 5250 | 71 | spring |  |
|  | 80000 | 11600 | 170 | summer |  |
|  | 96400 | 5500 | 67 | autumn |  |
|  | 102000 | 6230 | 72 | winter, pregnant |  |
|  | 97700 | 7330 | 88 | spring, pregnant |  |
|  | 79800 | 15000 | 220 | summer, lactating |  |
|  | 87800 | 11100 | 150 | metabolic chamber, simulated summer | Cameron et al. 1976 |
|  | 75500 | 1130 | 17 | metabolic chamber, simulated winter |  |
|  | 12200 | 1690 | 140 | nursing | McEwan and Whitehead 1971 |
| Caribou (H,M) Rangifer arcticus | 12500 | 1540 | 130 | nursing | McEwan and Whitehead 1971 |

*62*

Table A1, continued (Eutherian mammals, captive)

| | | | | |
|---|---|---|---|---|
| Guanaco (H,M) Lama guanacoe | 48000 | 1680 | 39 | | Macfarlane 1976 |
| Llama (H,M) Lama llamae | 103000 103000 103000 103000 | 2770 2090 2360 5440 | 32 24 27 62 | food water ad lib. food restricted water restricted in pasture | Rubsamen and Engelhardt 1975 |
| Llama (H,M) Lama peruana | 114000 | 2850 | 29 | | Macfarlane 1976 |
| Wildebeest (H,M and X) Connochaetes taurinus | 175000 | 9280 | 64 | | Macfarlane and Howard 1972 |
| Thompson's gazelle (H,X) Gazella dorcas | 175000 | 1470 | 90 | | Macfarlane and Howard 1972 |
| Oryx (H,M) Oryx beisa | 93000 93000 | 2790 11500 | 35 140 | minimum flux rate maximum flux rate | King 1979 |
| | 140000* | 8540 | 73 | | King et al. 1975 |
| | 136000 | 3940 | 35 | | Macfarlane and Howard 1972 |
| Moose (H,H) Alces alces | 186000 | 20600 | 134 | | Macfarlane and Howard 1972 |

63

Table A1, continued (Eutherian mammals, captive)

| GROUP Species (diet,habitat)[1] | Body mass, g | Water flux rate ml/day | % of predicted[2] | Conditions | Reference |
|---|---|---|---|---|---|
| Eland (H,M and X) *Taurotragus oryx* | 250000* | 25300 | 120 | | King et al. 1975, 1978 |
| | 204000 | 13500 | 81 | minimum flux rate | King 1979 |
| | 204000 | 36100 | 210 | maximum flux rate | |
| | 247000 | 19300 | 96 | | Macfarlane and Howard 1966a |
| Musk ox (H,M) *Ovibos moschatus* | 324000 | 11300 | 44 | | Macfarlane and Howard 1972 |
| Donkey (H,X) *Equus asinus* | 304000 | 21900 | 90 | | Macfarlane and Howard 1970 |
| | 160000 | 20200 | 150 | | Macfarlane et al. 1972 |
| | 375000 | 28000 | 94 | spring pasture | Macfarlane et al. 1974 |
| | 382000 | 17600 | 58 | autumn pasture | |
| | 370000* | 32600 | 110 | spring | Yousef 1971 |
| | 370000* | 47200 | 160 | summer | |
| | 98400 | 6050 | 72 | | Davis et al. 1978 |
| Horse (H,M) *Equus caballus* | 399000 | 21500 | 68 | | Richmond et al. 1962 |
| | 391000 | 54700 | 180 | arid conditions | Macfarlane 1968 |
| | 68800 | 14300 | 240 | colt, nursing | Doreau et al. 1980 |

64

Table A1, continued (Eutherian mammals, captive)

| | | | | |
|---|---|---|---|---|
| Cattle (H,M) | | | | |
| Bos indicus | 313000 | 32700 | 130 | various breeds | Springell 1968 |
| | 268000 | 19300 | 89 | low quality food | |
| | 330000 | 30400 | 120 | high quality food | |
| | 532000 | 65400 | 160 | tropics, wet season | Macfarlane and Howard 1972 |
| | 532000 | 64400 | 160 | | |
| | 223000 | 16600 | 91 | walk 8 km/d | Finch and King 1982 |
| | 213000 | 9270 | 53 | walk 8 km/d, restricted food | |
| | 208000 | 9960 | 58 | walk 16 km/d, restricted food | |
| | 194000 | 10400 | 65 | walk 16 km/d, restricted food and water | |
| | 192000 | 12400 | 78 | walk 16 km/d, restricted food and water | |
| | 300000 | 18900 | 78 | minimum flux rate | King 1979 |
| | 300000 | 53400 | 220 | maximum flux rate | |
| | 187000 | 25200 | 160 | dry tropics | Macfarlane 1965 |
| | 63000 | 12700 | 230 | nursing, tropics | Macfarlane et al. 1969 |
| | 284000 | 11400 | 50 | cool tropics | Macfarlane and Howard 1966b |
| Cattle (H,M) | | | | |
| Bos taurus | 244000 | 27800 | 140 | low quality food | Springell 1968 |
| | 193000 | 16800 | 106 | | |
| | 291000 | 32900 | 140 | high quality food | |
| | 278000 | 31700 | 140 | winter, dry pasture, water | Siebert and Macfarlane 1969 |
| | 314000 | 55000 | 220 | summer, dry pasture, water | |
| | 440000 | 85000 | 240 | winter, green pasture, water | |
| | 218000 | 24900 | 140 | humid tropics, pasture, water | |
| | 322000 | 54100 | 210 | tropics, wet season | |
| | 356000 | 76200 | 270 | tropics, wet season | |
| | 400000 | 68900 | 220 | tropics, wet season | |

Table A1, continued (Eutherian mammals, captive)

| GROUP Species (diet,habitat)[1] | Body mass, g | Water flux rate mL/day | Water flux rate % of predicted[2] | Conditions | Reference |
|---|---|---|---|---|---|
| Cattle (cont.) | 365000 | 35100 | 120 | pregnant, dry season | Siebert 1971 |
|  | 270000 | 54200 | 250 | lactating, wet season |  |
|  | 25000 | 4140 | 180 | calves nursing |  |
|  | 100000 | 9200 | 108 | calves nursing |  |
|  | 523000 | 73700 | 180 | cool tropics | Macfarlane and Howard 1966b |
|  | 46500 | 3660 | 88 | normal | Phillips and Knox 1969 |
|  | 32900 | 3590 | 120 | mild diarrhea |  |
|  | 29300 | 6560 | 250 | severe diarrhea |  |
|  | 382000 | 54600 | 180 | spring pasture | Macfarlane et al. 1974 |
|  | 470000 | 36200 | 98 | autumn pasture |  |
|  | 43500 | 4100 | 105 | 2 week old, nursing | Yates et al. 1971 |
|  | 48600 | 6840 | 160 | 4 week old, nursing |  |
|  | 55300 | 5510 | 110 | 6 week old, nursing |  |
|  | 58800 | 4890 | 95 | 8 week old, nursing |  |
|  | 46600 | 4580 | 110 | 2 week old, nursing |  |
|  | 52800 | 6020 | 130 | 4 week old, nursing |  |
|  | 59700 | 4600 | 88 | 6 week old, nursing |  |
|  | 65100 | 4730 | 83 | 8 week old, nursing |  |
|  | 51000 | 7990 | 180 | 2 week old, nursing |  |
|  | 59400 | 9110 | 170 | 4 week old, nursing |  |
|  | 67900 | 7690 | 130 | 6 week old, nursing |  |
|  | 73300 | 6530 | 103 | 8 week old, nursing |  |
|  | 50890 | 6430 | 140 | 2 week old, nursing |  |
|  | 64200 | 8370 | 150 | 4 week old, nursing |  |
|  | 74400 | 9160 | 140 | 6 week old, nursing |  |
|  | 90000 | 10400 | 130 | 8 week old, nursing |  |
|  | 110000 | 10400 | 110 | 11 week old, nursing |  |

Table A1, continued (Eutherian mammals, captive)

| | | | | |
|---|---|---|---|---|
| Cattle (cont.) | 400000* | 64800 | 200 | lactating | Aschbacher et al. 1965 |
| | 400000* | 36300 | 110 | nonlactating | |
| | 143000 | 19100 | 160 | wet season | Aggrey 1982 |
| | 139000 | 13400 | 110 | dry season | |
| | 75800 | 8800 | 130 | 49-54 day old nursing | Holleman et al. 1975 |
| | 106000 | 10600 | 120 | 72-82 day old nursing | |
| | 69000 | 8300 | 140 | 44 day old, nursing | Bailey and Lawson 1981 |
| | 100000 | 11700 | 140 | 79 day old, nursing and grazing | |
| | 130000 | 17500 | 160 | 114 day old, nursing and grazing | |
| | 160000 | 13500 | 101 | 142 day old, nursing and grazing | |
| | 175000 | 13700 | 95 | 163 day old, weaning | |
| | 350000* | 26300 | 94 | dry summer, water | Kamal 1982 |
| | 540000 | 29900 | 71 | | Knox et al. 1970 |
| | 75000 | 13100 | 200 | nursing, tropics | Macfarlane et al. 1969 |
| | 259000 | 39000 | 190 | | Baker et al. 1965 |
| | 185000 | 17300 | 110 | parasitized, diarrheic | |
| | 194000 | 28700 | 180 | dry tropics | Macfarlane et al. 1963 |
| | 500000* | 44000 | 110 | lactating | Argenzio et al. 1968 |
| | 596000 | 85000 | 180 | lactating | Black et al. 1964 |
| | 244000 | 34800 | 180 | nonlactating juvenile | |
| | 80000 | 13200 | 190 | calves | |
| | 402000 | 26900 | 84 | 18°C | Kamal et al. 1972 |
| | 407000 | 52800 | 160 | 32°C | |
| | 419000 | 31400 | 95 | | Maloiy and Hopcraft 1971 |

Table A1, continued (Eutherian mammals, captive)

| GROUP Species (diet,habitat)[1] | Body mass, g | Water flux rate mL/day | % of predicted[2] | Conditions | Reference |
|---|---|---|---|---|---|
| Cattle (cont.) | 172000 | 15600 | 109 | 18°C, 6 month old | Kamal and Johnson 1971 |
|  | 172000 | 26500 | 190 | 32°C, 6 month old |  |
|  | 530000 | 84100 | 200 |  | Potter et al. 1972 |
|  | 268000 | 31000 | 140 | winter | Ranjhan et al. 1982 |
|  | 317000 | 54800 | 220 | summer |  |
|  | 300000* | 29800 | 120 |  | Bird et al. 1980 |
|  | 300000* | 44500 | 180 |  |  |
|  | 382000 | 45100 | 150 | summer pasture | Wright 1982a |
|  | 353000 | 81000 | 290 | lactating | Wright 1982b |
|  | 435000 | 60600 | 180 | pregnant |  |
|  | 383000 | 45100 | 150 | steers, summer pasture |  |
|  | 127000 | 33000 | 310 | weaners, summer pasture |  |
| Cattle (H,M) Bos taurus X indicus | 294000 | 46100 | 190 | tropics, wet season | Siebert and Macfarlane 1969 |
|  | 523000 | 65400 | 160 | tropics, wet season |  |
| Cattle (H,M) Bubalus bubalis | 176000 | 33700 | 230 | tropics, wet season | Siebert and Macfarlane 1969 |
|  | 354000 | 70800 | 250 | tropics, wet season |  |
|  | 152000 | 23600 | 190 | nursing, tropics | Macfarlane et al. 1969 |
|  | 354000 | 75000 | 270 |  | Macfarlane and Howard 1972 |
|  | 355000* | 34100 | 120 | dry summer, water | Kamal 1982 |

Table A1, continued (Eutherian mammals, captive)

| | | | | |
|---|---|---|---|---|
| Cattle (cont.) | 287000 | 34800 | | 150 | Williams and Green 1982 |
| | 336000 | 25700 | 18°C | 96 | Kamal et al. 1972 |
| | 340000 | 46600 | 32°C | 170 | |
| Cattle (H,M) Bibos banteng | 215000 | 16800 | tropics, dry season | 95 | Siebert and Macfarlane 1969 |
| | 179000 | 27600 | tropics, dry season | 190 | |
| | 372000 | 47200 | tropics, wet season | 160 | |
| | 372000 | 49100 | | 170 | Macfarlane and Howard 1972 |
| | 51000 | 9280 | nursing, tropics | 210 | Macfarlane et al. 1969 |
| Camel (H,M and X) Camelus dromedarius | 400000 | 15400 | | 49 | Siebert and Macfarlane 1971 |
| | 541000 | 42100 | | 100 | |
| | 520000 | 31700 | | 78 | Macfarlane and Howard 1972 |
| | 716000 | 27600 | spring pasture | 50 | Macfarlane et al. 1974 |
| | 712000 | 15100 | autumn pasture | 28 | |
| | 500000* | 19550 | dry summer, water | 50 | Kamal 1982 |
| | 500000* | 19000 | minimum flux rate | 49 | King 1979 |
| | 500000* | 38000 | maximum flux rate | 97 | |
| | 337000 | 20600 | nonlactating | 76 | Macfarlane et al. 1963 |
| | 337000 | 31300 | lactating | 120 | |
| | 370000 | 12210 | | 41 | Macfarlane 1976 |
| | 500000 | 19300 | winter | 49 | Macfarlane and Siebert 1967 |
| | 500000 | 40000 | summer | 102 | |

69

Table A1, continued (Eutherian mammals, captive and field)

| GROUP Species (diet,habitat)[1] | Body mass, g | Water flux rate ml/day | % of predicted[2] | Conditions | Reference |
|---|---|---|---|---|---|
| Camel (cont.) | 420000 | 19800 | 60 | water ad lib. | Etzion et al. 1984 |
|  | 319000 | 6830 | 27 | dehydrated 20 days |  |
| Buffalo (H,M) |  |  |  |  |  |
| Syncerus caffer | 750000* | 81000 | 140 | minimum flux rate | King 1979 |
|  | 750000* | 152000 | 260 | maximum flux rate |  |
|  | 270000 | 31200 | 140 | winter | Ranjhan et al. 1982 |
|  | 330000 | 54200 | 200 | summer |  |
| In field |  |  |  |  |  |
| Bat (N,M) |  |  |  |  |  |
| Anoura caudifer | 11.5 | 13.4 | 560 | autumn | Helversen and Reyer 1984 |
| Bat (C,X) |  |  |  |  |  |
| Macrotus californicus | 13.3 | 1.8 | 67 | winter | Bell et al. 1986 |
|  | 12.6 | 3.0 | 120 | spring |  |
| House mouse (O,M) |  |  |  |  |  |
| Mus musculus | 13.9 | 3.25 | 120 | dry season | Morris and Bradshaw 1981 |
|  | 13.0 | 3.30 | 120 | autumn | Nagy and Morris (unpubl.) |
| Vole (O,M) |  |  |  |  |  |
| Clethrionomys rutilus | 15.0 | 3.6 | 120 | winter | Holleman et al. 1982 |
|  | 18.4 | 4.9 | 140 | spring |  |
|  | 15.8 | 5.7 | 180 | summer |  |
|  | 16.1 | 5.8 | 180 | autumn |  |
| Rock mouse (O,X) |  |  |  |  |  |
| Petromyscus collinus | 19 | 0.8 | 22 | before fog event (Namib) | Withers et al. 1980 |
|  | 19 | 1.4 | 39 | after fog |  |

Table A1, continued (Eutherian mammals, field)

| | | | | | |
|---|---|---|---|---|---|
| Pinyon mouse (G,M) | | | | | |
| Peromyscus truei | 20.2 | 3.22 | 85 | midsummer | Bradford 1974 |
| | 20.2 | 5.40 | 140 | late summer | |
| Pocket mouse (G,X) | | | | | |
| Perognathus formosus | 18* | 1.31 | 38 | summer | Mullen 1970 |
| | 18* | 2.23 | 64 | autumn | |
| Pocket mouse (G,X) | | | | | |
| Perognathus penicillatus | 14.9 | 1.7 | 57 | summer | Grenot and Serrano 1979 |
| | 18* | 1.78 | 51 | winter | Grubbs 1980 |
| | 18* | 1.31 | 38 | summer | |
| Pocket mouse (G,X) | | | | | |
| Perognathus fallax | 21.3 | 4.06 | 102 | spring | MacMillen and |
| | 21.3 | 2.61 | 66 | summer | Christopher 1975 |
| | 21.3 | 2.46 | 62 | autumn | |
| | 21.3 | 3.91 | 98 | winter | |
| Deer mouse (O,X) | | | | | |
| Peromyscus maniculatus | 25* | 5.28 | 120 | winter | Grubbs 1980 |
| | 15* | 3.56 | 120 | winter, juveniles | |
| Grey mouse (O,M) | | | | | |
| Pseudomys albocinereus | 19.2 | 3.92 | 107 | coastal habitat, dry season | Morris and Bradshaw 1981 |
| | 28.6 | 7.48 | 150 | coastal habitat, wet season | |
| | 19.4 | 3.01 | 82 | inland habitat, dry season | |
| | 29.3 | 6.51 | 130 | inland habitat, wet season | |
| | 32.6 | 7.76 | 140 | autumn | Nagy and Morris (unpubl.) |
| Kangaroo rat (G,X) | | | | | |
| Dipodomys merriami | 40.9 | 10.7 | 160 | spring | Mullen 1971 |
| | 31.7 | 4.2 | 76 | summer | |
| | 32.0 | 3.2 | 58 | autumn | |
| | 38.5 | 3.1 | 48 | winter | |

71

Table A1, continued (Eutherian mammals, field)

| GROUP Species (diet,habitat)[1] | Body mass, g | Water flux rate ml/day | % of predicted[2] | Conditions | Reference |
|---|---|---|---|---|---|
| Kangaroo rat (cont.) | 40.7 | 3.5 | 52 | summer | Grenot and Serrano 1979 |
|  | 40* | 3.70 | 56 | winter | Grubbs 1980 |
|  | 40* | 2.64 | 40 | summer |  |
|  | 37.3 | 6.59 | 105 | spring | MacMillen and Christopher 1975 |
|  | 37.3 | 4.19 | 67 | summer |  |
|  | 37.3 | 2.79 | 44 | autumn |  |
|  | 37.3 | 6.98 | 110 | winter |  |
| Kangaroo rat (O,X) Dipodomys microps | 55.1 | 10.7 | 120 | spring | Mullen 1971 |
|  | 55.0 | 6.2 | 72 | summer |  |
|  | 51.9 | 6.5 | 79 | autumn |  |
|  | 61.0 | 7.6 | 81 | winter |  |
| Kangaroo rat (G,X) Dipodomys nelsoni | 92.6 | 7.2 | 54 | summer | Grenot and Serrano 1979 |
| Bushy-tailed jird (O,X) Sekeetamys calurus | 41.2 | 5.89 | 86 | spring | Degen et al. 1986 |
| Spiny mouse (O,X) Acomys cahirinus | 38.3 | 5.06 | 79 | spring | Degen et al. 1986 |
|  | 49.0 | 5.45 | 69 |  | A. Shkolnik, in Morris and Bradshaw 1981 |
| Spiny mouse (O,X) Acomys russatus | 45.0 | 5.65 | 77 | spring | Degen et al. 1986 |

Table A1, continued (Eutherian mammals, field)

| | | | | | |
|---|---|---|---|---|---|
| Rock mouse (O,X) Aethomys namaquensis | 46 46 | 2.2 3.2 | 30 43 | before fog event (Namib) after fog | Withers et al. 1980 |
| Hopping mouse (O,X) Notomys mitchelli | 46.9 | 8.01 | 106 | winter | Morris and Bradshaw 1981 |
| Bush-rat (O,M) Rattus fuscipes | 63.5 | 16.9 | 107 | summer | Morris and Bradshaw 1981 |
| Sand rat (O,X) Meriones libycus | 59.6 | 8.9 | 97 | winter | Lachiver et al. 1978 |
| | 52.1 | 4.8 | 58 | spring | Chaouacha-Chekir et al. 1983 |
| | 85.0 | 10.8 | 88 | summer | Bradshaw et al. 1976b |
| Sand rat (O,X) Meriones shawii | 108 | 16.8 | 110 | summer | Bradshaw et al. 1976b |
| | 78.0 | 11.3 | 98 | winter | Lachiver et al. 1978 |
| | 39.7 | 7.4 | 110 | spring | Chaouacha-Chekir et al. 1983 |
| Sand rat (O,X) Psammomys obesus | 87.1 64.7 | 33.4 26.3 | 270 270 | seashore habitat, spring desert habitat, spring | Chaouacha-Chekir et al. 1983 |
| Vole (H,H and M) Arvicola terrestris | 52.6 84.4 93.6 | 45.7 62.9 78.3 | 550 510 590 | juveniles, summer adults, spring adults, summer | Grenot et al. 1984 |
| Ground squirrel (O,M) Spermophilus spilosoma | 101 39.6 | 27.5 8.0 | 190 120 | summer, lactating summer, juveniles | Grenot and Serrano 1979 |

Table A1, continued (Eutherian mammals, field)

| GROUP Species (diet,habitat)[1] | Body mass, g | Water flux rate ml/day | % of predicted[2] | Conditions | Reference |
|---|---|---|---|---|---|
| Ground squirrel (O,X) *Ammospermophilus leucurus* | 96.1 | 8.7 | 64 | winter | Karasov 1983a |
| | 90.0 | 18.0 | 140 | spring | |
| | 79.9 | 12.0 | 102 | summer | |
| | 82.1 | 9.0 | 75 | autumn | |
| Pocket gopher (H,M) *Thomomys bottae* | 99 | 26 | 190 | summer | Gettinger 1984 |
| | 108 | 27 | 180 | winter | |
| | 104 | 53 | 365 | spring | |
| Rock rat (O,X) *Petromus typicus* | 130 | 4.4 | 25 | before fog event (Namib) | Withers et al. 1980 |
| | 130 | 5.1 | 29 | after fog | |
| Wood rat (O,X) *Neotoma albigula* | 144 | 19.9 | 105 | summer | Grenot and Serrano 1979 |
| Rabbit (H,M) *Oryctolagus cuniculus* | 1270 | 291 | 260 | water available, low-Na plants | Green et al. 1978 |
| | 1630 | 265 | 190 | no water, low-Na plants | |
| | 1350 | 376 | 320 | no water, high-Na plants | |
| | 1290 | 317 | 280 | water available, high-Na plants | |
| Rabbit (H,X) *Oryctolagus cuniculus* | 1410 | 302 | 250 | winter, wet pasture, arid | Richards 1979 |
| | 1540 | 262 | 200 | spring intermediate pasture | |
| | 1410 | 175 | 140 | summer, intermediate pasture | |
| | 1710 | 136 | 95 | winter, dry pasture | |
| | 1500 | 83 | 64 | summer, dry pasture | |
| | 1510 | 286 | 220 | summer, wet pasture | |
| | 1070 | 223 | 230 | subadults, spring | |
| | 432 | 125 | 270 | juveniles, spring | |

Table A1, continued (Eutherian mammals, field)

| | | | | |
|---|---|---|---|---|
| Jackrabbit (H,X) | | | | |
| Lepus californicus | 1800 | 580 | 390 | spring |
| | 1800 | 82.4 | 55 | summer |
| | 1800 | 67.3 | 45 | winter |
| | | | | |
| Kit fox (O,X) | | | | |
| Vulpes macrotis | 1850 | 137 | 90 | summer and winter |
| | | | | |
| Three-toed sloth (H,M) | | | | |
| Bradypus variegatus | 3830 | 154 | 56 | females with young |
| | 4220 | 125 | 42 | females without young |
| | 4450 | 181 | 58 | males |
| | | | | |
| Howler monkey (H,M) | | | | |
| Alouatta palliata | 6500 | 767 | 180 | dry season |
| | | | | |
| Peccary (O,X) | | | | |
| Dicotyles tajacu | 19800 | 1360 | 130 | summer |
| | 19100 | 1170 | 110 | winter |
| | | | | |
| Hyena (C,M) | | | | |
| Crocuta crocuta | 54300 | 3500 | 140 | winter, spring |
| | | | | |
| Fur seal (C,SW) | | | | |
| Callorhinus ursinus | 6130 | 345 | 85 | pups, nursing |
| | 7370 | 181 | 38 | female pups, fasting |
| | 8990 | 166 | 30 | male pups, fasting |
| | 42700 | 487 | 24 | lactating females, fasting |
| | 44100 | 520 | 25 | nonlactating females, fasting |
| | 30900 | 5690 | 370 | lactating females, foraging |
| | | | | |
| Deer (H,M) | | | | |
| Odocoileus hemionus | 40000 | 4550 | 240 | females, spring |
| | 67100 | 8050 | 280 | males, spring |

References (right column):
Nagy et al. 1976
Golightly and Ohmart 1984
Nagy and Montgomery 1980
Nagy and Milton 1979b
Zervanos and Day 1977
Green et al. 1984
Costa and Gentry 1986
Nagy and Jacobsen (unpubl.)

75

Table A1, continued (Eutherian mammals, field and Marsupial mammals, captive)

| GROUP Species (diet,habitat)[1] | Body mass, g | Water flux rate mL/day | % of predicted[2] | Conditions | Reference |
|---|---|---|---|---|---|
| Sea lion (C,SW) Zalophus californianus | 75200 82800 | 7640 7620 | 240 220 | lactating, 1983 lactating, 1984 | Costa 1984 |
| Lion (C,M) Panthera leo | 140000 | 7540 | 140 | winter, spring | Green et al. 1984 |
| Elephant seal (C,SW) Mirounga angustirostris | 134000 | 716 | 14 | pup, weaned, fasting on beach | Ortiz et al. 1978 |
|  | 425000 95000 | 1870 1870 | 14 49 | lactating females pups, nursing | Costa et al. (in press) |
|  | 86000 | 2690 | 76 | pups, nursing | Ortiz et al. 1984 |
| Water buffalo (H,M and H) Bubalus bubalis | 429000 | 45600 | 350 | dry season | Williams and Ridpath 1982 |

MARSUPIAL MAMMALS

In captivity

| Dunnart (C,M) Sminthopsis crassicaudata | 16.0 15.3 14.9 14.2 |  | 9.1 7.4 7.5 5.0 | 200 170 170 120 | crickets and water ad lib crickets, no water mealworms and water ad lib. mealworms, no water | Morton 1980 |
|---|---|---|---|---|---|---|
|  | 17 |  | 8.2 | 170 |  | Macfarlane 1975 |
|  | 19 |  | 8.8 | 170 |  | Kennedy and Macfarlane 1971 |

Table A1, continued (Marsupial mammals, captive)

| | | | | |
|---|---|---|---|---|
| Marsupial mouse (C,X) | | | | |
| Antechinomys laniger | 18 | 6.7 | 130 | | Macfarlane 1975 |
| Mulgara (C,X) | | | | |
| Dasycercus cristicauda | 86 | 11.5 | 68 | | Kennedy and Macfarlane 1971 |
| | 87 | 11.3 | 66 | | Macfarlane 1975 |
| Kowari (C,X) | | | | |
| Dasyuroides byrnei | 124 | 14.3 | 64 | | Macfarlane 1975 |
| | 127 | 16.8 | 73 | water ad lib. | Haines et al. 1974 |
| | 121 | 10.0 | 45 | water restricted | |
| | 130 | 15.6 | 67 | | Macfarlane and Howard 1972 |
| Bandicoot (C,M) | | | | |
| Macrotis lagotis | 1080 | 48.1 | 40 | water ad lib. | Hulbert and Dawson 1974 |
| | 943 | 29.1 | 27 | water restricted | |
| Bandicoot (C,M) | | | | |
| Perameles nasuta | 972 | 68.3 | 62 | water ad lib. | Hulbert and Dawson 1974 |
| | 837 | 37.9 | 39 | water restricted | |
| Bandicoot (C,M) | | | | |
| Isoodon macrourus | 1470 | 131 | 87 | water ad lib | Hulbert and Dawson 1974 |
| | 1180 | 39.7 | 31 | water restricted | |
| | 1170 | 149 | 120 | lactating | |
| Native cat (C,M) | | | | |
| Dasyurus viverrinus | 1340 | 162 | 120 | | Green and Eberhard 1979 |

Table A1, continued (Marsupial mammals, captive and field).

| GROUP Species (diet,habitat)[1] | Body mass, g | Water flux rate mL/day | % of predicted[2] | Conditions | Reference |
|---|---|---|---|---|---|
| Potoroo (H,M) *Potorous tridactylus* | 1400 | 137 | 94 | | Denny and Dawson 1975 |
| Rock wallaby (H,X and M) *Petrogale inornata* | 3410 3510 | 380 284 | 131 96 | water ad lib. green food, no water | Kennedy and Heinsohn 1974 |
| Pademelon wallaby (H,M) *Thylogale thetis* | 4800 4800 | 775 1790 | 210 480 | distilled water saline water | Hume and Dunning 1979 |
| Tasmanian devil (C,M) *Sarcophilus harrisii* | 5250 | 392 | 97 | | Nicol 1978 |
| | 3840 | 383 | 120 | | Green and Eberhard 1979 |
| Tammar wallaby (H,M) *Macropus eugenii* | 6500 | 291 | 61 | | Denny and Dawson 1975 |
| | 4900 4900 | 681 1080 | 180 280 | distilled water saline water | Hume and Dunning 1979 |
| Grey kangaroo (H,M) *Macropus giganteus* | 22100 | 937 | 77 | | Denny and Dawson 1975 |
| Red kangaroo (H,X) *Macropus rufus* | 23400 | 1430 | 110 | | Denny and Dawson 1975 |
| | 22000 | 1150 | 95 | | Denny and Dawson 1973 |
| | 22000* | 1940 | 160 | dry tropics | Macfarlane et al. 1963 |

Table A1, continued (Marsupial mammals, captive and field)

| | | | | | |
|---|---|---|---|---|---|
| Euro kangaroo (H,M) | | | | | |
| Macropus robustus | 31000 | 1850 | 120 | | Denny and Dawson 1975 |
| In field | | | | | |
| Dunnart (C,M) | | | | | |
| Sminthopsis crassicaudata | 20.4 | 22.4 | 150 | summer | Morton 1980 |
| | 10.7 | 12.4 | 120 | autumn | |
| | 13.3 | 19.8 | 170 | winter | |
| | 15.3 | 20.6 | 160 | spring | |
| | 6.1 | 3.46 | 47 | juveniles, spring | Nagy et al. (unpubl.) |
| | 16.6 | 13.4 | 99 | adults, spring | |
| Brown antechinus (C,M) | | | | | |
| Antechinus stuartii | 25.7 | 13.9 | 79 | winter, prebreeding | Nagy et al. 1978 |
| | 25.7 | 18.9 | 108 | winter, breeding | |
| Marsupial mouse (C,M) | | | | | |
| Antechinus swainsonii | 26.3 | 15.0 | 84 | juvenile females, spring | Lee and Nagy (unpubl.) |
| | 32.1 | 18.1 | 90 | juvenile males, spring | |
| | 47.4 | 23.1 | 91 | females, spring | |
| | 54.2 | 72.5 | 260 | females, lactating | |
| | 52.5 | 36.4 | 130 | females, mating season | |
| | 72.7 | 50.4 | 150 | males, mating season | |
| Sugar glider (O,M) | | | | | |
| Petaurus breviceps | 135 | 40.6 | 85 | males, spring | Nagy and Suckling 1985 |
| | 112 | 21.7 | 51 | females, spring | |
| Leadbeater's possum (O,M) | | | | | |
| Gymnobelideus leadbeateri | 133 | 42.8 | 91 | males, winter | Smith et al. 1982 |
| | 135 | 32.6 | 68 | males, spring | |
| | 95 | 44.5 | 120 | females, winter | |
| | 118 | 37.9 | 86 | females, spring | |

Table A1, continued (Marsupial mammals, field)

| GROUP Species (diet,habitat)[1] | Body mass, g | Water flux rate ml/day | % of predicted[2] | Conditions | Reference |
|---|---|---|---|---|---|
| Ringtail possum (H,M) |  |  |  |  |  |
| Pseudocheirus peregrinus | 278 | 43.6 | 59 | juveniles, spring | Nagy et al. (unpubl.) |
|  | 717 | 106 | 81 | adults, spring |  |
| Greater glider (H,M) |  |  |  |  |  |
| Petauroides volans | 934 | 80.4 | 53 | females, winter | Foley and Nagy (unpubl.) |
|  | 1042 | 94.8 | 58 | males, winter |  |
| Native cat (C,M) |  |  |  |  |  |
| Dasyurus viverrinus | 1120 | 202 | 120 | summer | Green and Eberhard 1983 |
|  | 1270 | 236 | 130 | autumn |  |
|  | 920 | 261 | 170 | winter |  |
|  | 1260 | 265 | 150 | spring, males |  |
|  | 984 | 222 | 140 | spring, females, early lactation |  |
|  | 984 | 332 | 210 | spring, females, late lactation |  |
| Bandicoot (C,M) |  |  |  |  |  |
| Isoodon macrourus | 1410 | 354 | 180 | water available | Hulbert and Gordon 1972 |
|  | 1390 | 355 | 180 | no water available |  |
| Bandicoot (C,M) |  |  |  |  |  |
| Isoodon obesulus | 1230 | 109 | 61 | adults, autumn | Nagy et al. (unpubl.) |
| Brush-tailed possum (H,M) |  |  |  |  |  |
| Trichosurus vulpecula | 1520 | 136 | 67 | nonlactating | Kennedy and Heinsohn 1974 |
|  | 1590 | 165 | 79 | lactating |  |
| Quokka wallaby (H,M) |  |  |  |  |  |
| Setonix brachyurus | 1510 | 72 | 35 | juveniles, summer | Nagy et al. (unpubl.) |
|  | 2470 | 116 | 42 | adults, summer |  |
|  | 3390 | 357 | 108 | males, mild period | Kitchener 1970 |
|  | 2630 | 271 | 95 | females, hot period |  |

Table A1, continued (Marsupial mammals, field)

| | | | | | |
|---|---|---|---|---|---|
| Rock wallaby (H,X and M) | | | | | |
| Petrogale inornata | 3180 | 257 | 81 | nonlactating | Kennedy and Heinsohn 1974 |
| | 3140 | 323 | 102 | lactating | |
| Tammar wallaby (H,M) | | | | | |
| Macropus eugenii | 4560 | 262 | 66 | adults, summer | Nagy et al. (unpubl.) |
| | 5000* | 349 | 83 | summer, no water | Bakker et al. 1982 |
| | 5000* | 435 | 104 | autumn, no water | |
| | 5000* | 664 | 160 | winter, no water | |
| | 5000* | 336 | 80 | summer, water available | |
| | 5000* | 329 | 79 | autumn, water available | |
| | 5000* | 630 | 150 | winter, water available | |
| Pademelon wallaby (H,M) | | | | | |
| Thylogale billardieri | 5450 | 621 | 140 | adults, spring | Nagy and Sanson (unpubl.) |
| Koala (H,M) | | | | | |
| Phascolarctos cinereus | 10800 | 475 | 71 | males, spring | Nagy and Martin 1985 |
| | 7800 | 358 | 65 | females, spring | |
| | 6910 | 843 | 170 | summer | Degabriele et al. 1978 |
| Red kangaroo (H,X) | | | | | |
| Macropus rufus | 21800 | 861 | 85 | summer | Dawson et al. 1975 |
| Euro kangaroo (H,M) | | | | | |
| Macropus robustus | 28100 | 1110 | 94 | summer | Dawson et al. 1975 |
| | 27000 | 1350 | 120 | spring | Denny and Dawson 1973 |
| Grey kangaroo (H,M) | | | | | |
| Macropus giganteus | 27000 | 1570 | 140 | females, spring | Nagy and Sanson (unpubl.) |
| | 61900 | 3840 | 200 | males, spring | |

Table A1, continued (Monotreme mammal, field and Birds, captive)

| GROUP Species (diet, habitat)[1] | Body mass, g | Water flux rate ml/day | % of predicted[2] | Conditions | Reference |
|---|---|---|---|---|---|
| **MONOTREME MAMMAL** | | | | | |
| *In field* | | | | | |
| Platypus (C,FW) | | | | | |
| Ornithorhynchus anatinus | 856 | 478 | | summer | Hulbert and Grant 1983 |
| **BIRDS** | | | | | |
| *In captivity* | | | | | |
| Black-rumped waxbill (G,M) | | | | | |
| Estrilda troglodytes | 6.7 | 1.49 | 46 | aviary, spring, no water | Weathers and Nagy 1984 |
| Zebra finch (G,M and X) | | | | | |
| Taeniopygia guttata | 13.3 | 2.80 | 53 | dehydrated | Skadhauge and Bradshaw 1974 |
| | 13.3 | 7.34 | 140 | normally hydrated | |
| | 12.5 | 6.4 | 130 | 13°C | S. Ambrose, in Rooke et al. 1983 |
| | 12.5 | 6.1 | 120 | 24°C | |
| | 12.5 | 6.8 | 130 | 36°C | |
| Savannah sparrow (O,H) | | | | | |
| Passerculus sandwichensis | 16.1 | 7.46 | 120 | | Williams and Nagy 1984a |
| Song sparrow (O,M) | | | | | |
| Melospiza melodia | 20* | 8.14 | 120 | aviary, July | Stephenson and Minnich 1974 |
| | 20* | 8.42 | 120 | aviary, August | |
| | 20* | 2.56 | 37 | aviary, January | |
| | 20.6 | 6.9 | 97 | | Williams 1985 |

Table A1, continued (Birds, captive)

| Species | | | | Reference |
|---|---|---|---|---|
| White-throated sparrow (O,M) *Zonotrichia albicolis* | 23.2 | 8.3 | 107 | | Williams 1985 |
| House sparrow (O,M) *Passer domesticus* | 23.4 | 9.3 | 120 | | Williams 1985 |
| Semipalmated sandpiper (C,H) *Calidris pusillus* | 24.3<br>22.5 | 13.2<br>12.7 | 160<br>170 | 25°C<br>40°C | Purdue and Haines 1977 |
| Snowy plover (C,H) *Charadrius alexandrinus* | 33.7<br>32.2<br>34.0 | 14.6<br>15.6<br>13.3 | 150<br>160<br>130 | 25°C<br>40°C<br>25°C, saline water | Purdue and Haines 1977 |
| Starling (O,M) *Sturnus vulgaris* | 69.8 | 39.9 | 240 | | Williams 1985 |
| Killdeer (C,M) *Charadrius vociferus* | 71.1 | 32.9 | 200 | 40°C | Purdue and Haines 1977 |
| Quail (G,M) *Coturnix coturnix* | 117<br>105 | 23.5<br>23.1 | 98<br>104 | males<br>females | Chapman and McFarland 1971 |
| Burrowing owl (C,M and X) *Speotyto cunicularia* | 140 | 10.2 | 38 | | Chapman and McFarland 1971 |
| Gambel's quail (O,X) *Callipepla gambelii* | 141 | 15.7 | 58 | natural diet, water | Goldstein and Nagy 1985 |
| Sand partridge (O,X) *Ammoperdix heyi* | 164<br>159 | 12.6<br>14.1 | 42<br>48 | | Pinshow et al. 1983<br>Degen et al. 1981 |

Table A1, continued (Birds, captive)

| GROUP Species (diet, habitat)[1] | Body mass, g | Water flux rate ml/day | % of predicted[2] | Conditions | Reference |
|---|---|---|---|---|---|
| Partridge (O,M) *Perdix perdix* | 161 | 41.1 | 140 | outdoor enclosure | Pinet et al. 1982 |
| Roadrunner (C,X and M) *Geococcyx californianus* | 306 | 34.2 | 73 | outdoor cage, winter | Ohmart et al. 1970 |
|  | 288 | 22.5 | 50 | indoor cage, food, no water |  |
| Petz conure (O,M) *Aratinga canicular* | 313 | 17.8 | 38 |  | Chapman and McFarland 1971 |
| Chukar partridge (O,M and X) *Alectoris chukar* | 354 | 32.0 | 62 | green and dry food, water | Degen et al. 1982 |
|  | 389 | 27.5 | 50 | no green food, water ad lib. |  |
|  | 404 | 23.2 | 41 | dry food, water, indoors | Degen et al. 1981 |
|  | 420 | 30.4 | 52 | dry food, water, outdoors | Degen et al. 1983b |
|  | 436 | 21.9 | 37 | dry and green food, no water |  |
|  | 405 | 24.2 | 43 | dry and green food, water |  |
|  | 318 | 32.8 | 44 | green food only |  |
|  | 416 | 28.6 | 50 |  | Pinshow et al. 1983 |
| Galah (O,M) *Cacatua roseicapilla* | 406 | 44.7 | 79 | 13°C | S. Ambrose, in Rooke et al. 1983 |
|  | 406 | 36.5 | 64 | 24°C |  |
|  | 406 | 44.7 | 79 | 36°C |  |
| Glaucous-winged gull (C,H) *Larus glaucescens* | 787 | 50.4 | 56 |  | Walter and Hughes, 1978 |

*84*

Table A1, continued (Birds, captive)

| | | | | |
|---|---|---|---|---|
| Mallard (O,H) *Anas platyhynchos* | 1190 | 158 | 130 | | M.R. Hughes, in Pinshow et al. 1983 |
| Fish eagle (C,H) *Gypohierax angolensis* | 1590 | 134 | 92 | | Chapman and McFarland 1971 |
| Chicken (G,M) *Gallus gallus* | 2150 | 112 | 62 | hens, 7 year old | Lopez et al. 1973a |
| | 2680 | 235 | 110 | pullets | |
| | 2600 | 159 | 77 | roosters | Chapman and Black 1967 |
| | 1660 | 208 | 140 | laying hens | |
| | 2680 | 235 | 110 | pullets, 5 month old | Lopez et al. 1973b |
| | 2780 | 342 | 160 | pullets, 5 month old, low vitamin A | |
| | 5090 | 326 | 99 | males, 24°C maximum temperature | Chapman and Mihai 1972 |
| | 3490 | 255 | 101 | females, 24°C maximum temperature | |
| | 4900 | 353 | 110 | males, 32°C maximum temperature | |
| | 3440 | 420 | 170 | females, 32°C maximum temperature | |
| | 4640 | 278 | 90 | males, 43°C maximum temperature | |
| | 3530 | 441 | 170 | females, 43°C maximum temperature | |
| Domestic duck (O,H) *Anas platyrhynchos* | 3060 | 903 | 390 | adrenalectomized | Thomas and Phillips 1975 |
| | 3100 | 1960 | 840 | sham-operated | |
| | 3130 | 1760 | 750 | control (intact) | |
| Canada goose (O,H) *Branta canadensis* | 3650 | 278 | 107 | | M.R. Hughes, in Pinshow et al. 1983 |
| Domestic goose (O,H) *Anser anser* | 6250 | 796 | 210 | | Le Maho, in Pinshow et al. 1983 |

Table A1, continued (Birds, captive and field)

| GROUP Species (diet, habitat)[1] | Body mass, g | Water flux rate ml/day | % of predicted[2] | Conditions | Reference |
|---|---|---|---|---|---|
| Emu (O,M and X) | | | | | |
| *Dromaius novae-hollandiae* | 35400 | 2000 | 160 | | E. Skadhauge, in Pinshow et al. 1983 |
| | 32700 | 1950 | 160 | hydrated | E. Shadhauge, in Rooke et al. 1983 |
| | 30400 | 608 | 53 | dehydrated | |
| | 38800 | 1740 | 130 | hydrated | Dawson et al. 1983 |
| | 34200 | 274 | 22 | dehydrated | |
| | 30200 | 1640 | 140 | adults, summer | Dawson et al. 1984 |
| | 6600 | 1270 | 320 | chicks, summer | |
| | 34500 | 1820 | 150 | adults, winter | |
| | 2500 | 763 | 380 | chicks, late spring | |
| | 34800 | 183 | 14 | adult, on nest | |
| | 40200 | 2690 | 190 | adults, off nest | |
| Ostrich (O,X and M) | | | | | |
| *Struthio camelus* | 95400 | 8230 | 330 | water ad lib. | Withers 1983 |
| | 80700 | 550 | 25 | water restricted | |

In field

| Anna's hummingbird (N,M) | | | | | |
|---|---|---|---|---|---|
| *Calypte anna* | 4.5 | 5.1 | 130 | summer | Powers and Nagy (unpubl.) |
| Silvereye (O,M) | | | | | |
| *Zosterops lateralis* | 9.1 | 5.53 | 87 | autumn | Rooke et al. (unpubl.) |
| | 9.7 | 11.6 | 180 | spring, 1980 | |
| | 9.6 | 12.8 | 190 | spring, 1983 | |

Table A1, continued (Birds, field)

| Species | | | | |
|---|---|---|---|---|
| Silvereye (cont.) | 9.5 | 13.7 | 210 | spring, coastal heath | Rooke et al. 1983 |
| | 10.2 | 22.5 | 330 | summer, coastal heath | |
| | 9.2 | 6.0 | 94 | summer, vineyard | |
| White-browed scrub-wren (C,M) *Sericornis frontalis* | 11.4 | 9.35 | 130 | | S. Ambrose, in Rooke et al. 1983 |
| White-fronted chat (C,M and X) *Epthianura albifrons* | 12.4 | 7.69 | 98 | | C. Williams, in Rooke et al. 1983 |
| House finch (O,M) *Carpodacus mexicanus* | 15 | 5.2 | 58 | nestlings | Gettinger et al. 1985 |
| Savannah sparrow (O,M) *Passerculus sandwichensis* | 17.0 | 13.0 | 130 | females, feeding young | Williams and Nagy 1985a |
| | 19.1 | 16.4 | 150 | males, territorial | Williams and Nagy 1984b |
| | 2.9 | 2.5 | 87 | 0.5 day old nestlings | Williams and Nagy 1985b |
| | 9.6 | 6.2 | 94 | 4.5 day old nestlings | |
| | 14.6 | 9.8 | 110 | 6.5 day old nestlings | |
| Song sparrow (O,M) *Melospiza melodia* | 20* | 10.6 | 97 | July | Stephenson and Minnich 1974 |
| | 20* | 11.5 | 105 | August | |
| Phainopepla (O,X and M) *Phainopepla nitens* | 22.7 | 21.5 | 180 | desert and aviary, spring | Weathers and Nagy 1980 |
| Red-winged blackbird (C,H) *Agelaius phoeniceus* | 30.2 | 36.8 | 250 | nestlings, daylight | Fiala and Congdon 1983 |
| | 30.2 | 12.1 | 83 | nestlings, at night | |
| Wilson's storm petrel (C,SW) *Oceanites oceanicus* | 42.3 | 3.05 | 17 | incubating | Obst et al. 1987 |
| | 42.2 | 25.1 | 140 | foraging | |

Table A1, continued (Birds, field)

| GROUP Species (diet,habitat)[1] | Body mass, g | Water flux rate mL/day | % of predicted[2] | Conditions | Reference |
|---|---|---|---|---|---|
| Mockingbird (O,M) Mimus polyglottos | 47.6 | 42.1 | 210 | breeding season | Utter 1971 |
| Purple martin (C,M) Progne subis | 47.7 50.3 | 33.0 30.7 | 160 150 | females, feeding nestlings males | Utter and LeFebvre 1973 |
| Leach's storm petrel (C,SW) Oceanodroma leucorhoa | 48 48 | 2.94 21.8 | 15 110 | incubating eggs away from nest | Ricklefs et al. 1986 |
| Starling (O,M) Sturnus vulgaris | 85.0 78.7 74.1 76.9 | 62.7 77.7 79.7 80.3 | 210 270 290 290 | females, incubating stage females, early nestling stage females, middle nestling stage males, middle nestling stage | Ricklefs and Williams 1984 |
| Gambel's quail (O,X) Callipepla gambellii | 143 | 17.7 | 41 | late summer | Goldstein and Nagy 1985 |
| Sand partridge (O,X) Ammoperdix heyi | 173 156 206 209 | 21.2 17.9 24.3 17.8 | 43 39 44 32 | summer summer spring autumn | Degen et al. 1983a Kam et al. 1987 |
| Kittiwake (C,SW) Rissa tridactyla | 386 | 124 | 140 | breeding | Gabrielsen et al. 1987 |

Table A1, continued (Birds, field)

| | | | | |
|---|---|---|---|---|
| Chukar partridge (O,M and X) | | | | |
| Alectoris chukar | 515 | 108 | 103 | winter, males, 1981 | Alkon et al. 1985 |
| | 429 | 99.6 | 108 | winter, females, 1981 | |
| | 490 | 70.6 | 70 | winter, males, 1982 | Alkon et al. 1982 |
| | 413 | 75.1 | 84 | winter, females, 1982 | |
| | 446 | 44.1 | 47 | winter, before rains, 1982 | |
| | 422 | 43.4 | 48 | winter, during rains, 1982 | |
| | 447 | 147 | 160 | winter, after rains, 1982 | |
| | 445 | 44.7 | 47 | summer | Degen et al. 1983a |
| | 333 | 40.3 | 52 | summer | Kam et al. 1987 |
| | 432 | 41.5 | 45 | spring | |
| | 419 | 45.1 | 50 | autumn | |
| Laysan albatross (C,SW) | | | | | |
| Diomedea immutabilis | 3070 | 58.3 | 16 | incubating | Ellis et al. (unpubl.) |
| | 3060 | 551 | 150 | foraging | |
| Jackass penguin (C,SW) | | | | | |
| Spheniscus demersus | 3170 | 288 | 78 | breeding season | Nagy et al. 1984b |
| Giant petrel (C;SW) | | | | | |
| Macronectes giganteus | 3580 | 469 | 120 | females, breeding | Obst and Nagy (unpubl.) |
| | 4510 | 586 | 120 | males, breeding | |
| Macaroni penguin (C,SW) | | | | | |
| Eudyptes chrysolophus | 3600 | 662 | 160 | foraging at sea | Davis et al. 1983 |
| | 3750 | 49 | 12 | fasting in rookery | |
| Grey-headed albatross (C,SW) | | | | | |
| Diomedea chrysostoma | 3650 | 1040 | 250 | summer | Costa and Prince 1987 |
| Adelie penguin (C,SW) | | | | | |
| Pygoscelis adeliae | 3870 | 565 | 130 | breeding | Nagy and Obst (unpubl.) |

Table A1, continued (Birds, field and Reptiles, captive)

| GROUP Species (diet, habitat)[1] | Body mass, g | Water flux rate mL/day | % of predicted[2] | Conditions | Reference |
|---|---|---|---|---|---|
| Gentoo penguin (C,SW) Pygoscelis papua | 6200 | 961 | 160 | foraging at sea | Davis et al. 1983 |
| Wandering albatross (C,SW) Diomedea exulans | 7360 9440 | 1330 1530 | 200 200 | females, breeding males, breeding | Adams et al. 1986 |
| King penguin (C,SW) Aptenodytes patagonica | 13000 | 2200 | 220 | foraging at sea, breeding | Kooyman et al. 1982 |

REPTILES

In captivity

| Skink (C,M) Lampropholis guichenoti | 1.23 1.07 1.37 1.10 | 0.126 0.127 0.179 0.160 | 53 59 70 73 | eat 1 fly larva/day eat 2 fly larvae/day eat 3 fly larvae/day eat 4 fly larvae/day | Morrison and Gallagher 1984 |
|---|---|---|---|---|---|
| Sea snake (C,SW) Hydrophis cyanocinctus | 20 | 3.60 | 200 | in seawater, starved | Dunson 1978 |
| Green lacerta (C,M) Lacerta viridis | 25.5 | 1.91 | 89 | outdoor enclosure, summer | Bradshaw et al. 1987 |
| Sea snake (C,SW) Lapemis hardwickii | 39 | 4.02 | 140 | in seawater, starved | Dunson 1978 |
| Salt marsh water snake (C, SW and FW) Nerodia fasciata clarki | 40 | 1.15 | 39 | in seawater, starved | Dunson 1978 |

Table A1, continued (Reptiles, captive)

| Species | | | | |
|---|---|---|---|---|
| Common garter snake (C,M) *Thamnophis sirtalis* | 41 | 17.8 | 590 | in seawater, starved | Dunson 1978 |
| Sea snake (C,SW) *Hydrophis ornatus* | 42 | 10.9 | 350 | in seawater, starved | Dunson 1978 |
| Queen snake (C,FW) *Regina septemvittata* | 44 | 50.9 | 1600 | in seawater, starved | Dunson 1978 |
| Yellow-bellied sea snake (C,SW) *Pelamis platurus* | 46 | 2.65 | 81 | in seawater, starved | Dunson and Robinson 1976 |
| File snake (C,SW and FW) *Acrochordus granulatus* | 54 | 6.09 | 170 | in seawater, starved | Dunson 1978 |
| Asp viper (C,M) *Vipera aspis* | 67.2 | 0.97 | 22 | outdoor enclosure, summer | Bradshaw et al. 1987 |
| Mangrove snake (C,SW and FW) *Nerodia valida* | 68 | 2.94 | 67 | in seawater, starved | Dunson 1978 |
| Common water snake (C,FW) *Nerodia sipedon* | 72 | 23.0 | 510 | in seawater, starved | Dunson 1978 |
| Sea snake (C,SW) *Laticauda semifasciata* | 77 | 2.96 | 62 | in seawater, starved | Dunson 1978 |
| Dog-faced water snake (C,SW and FW) *Cereberus rhynchops* | 78 | 3.74 | 78 | in seawater, starved | Dunson 1978 |
| Sea snake (C,SW) *Hydrophis inornatus* | 85 | 24.3 | 470 | in seawater, starved | Dunson 1978 |
| Sea snake (C,SW) *Laticauda laticaudata* | 92 | 3.75 | 69 | in seawater, starved | Dunson 1978 |

Table A1, continued (Reptiles, captive)

| GROUP Species (diet, habitat)[1] | Body mass, g | Water flux rate ml/day | % of predicted[2] | Conditions | Reference |
|---|---|---|---|---|---|
| Chuckwalla (H,X) Sauromalus obesus | 92.8 | 1.68 | 31 | force-fed spring diet, no water | Nagy 1972 |
| Sea snake (C,SW) Hydrophis belcheri | 100 | 13.0 | 230 | in seawater, starved | Dunson 1978 |
| Florida banded water snake (C,FW) Nerodia fasciata pictiventris | 109 | 12.8 | 210 | in seawater, starved | Dunson 1980 |
| Sea snake (C,SW) Laticauda colubrina | 136 | 2.94 | 62 | in seawater, starved | Dunson 1978 |
| Saltwater crocodile (C,SW and FW) Crocodylus porosus | 206 161 185 153 | 12.9 35.6 39.7 31.9 | 130 440 440 410 | in seawater, SW acclimated, starved in seawater, FW acclimated, starved in seawater, FW acclimated, starved as above, salt glands sealed | Taplin 1985 |
| Painted turtle (H,FW) Chrysemys picta | 245 | 42.3 | 380 | in seawater | Robinson and Dunson 1976 |
| Spiny-tailed agamid (H,X) Uromastix acanthinurus | 385 377 392 360 474 481 377 | 6.39 2.86 2.08 1.48 5.55 2.26 1.32 | 42 19 13 10 31 13 9 | seminatural, autumn, moist food seminatural, autumn, dry food seminatural, autumn, fasting seminatural, autumn, underground seminatural, spring, moist food seminatural, spring, dry food seminatural, spring, underground | Lemire et al. 1979 |

Table A1, continued (Reptiles, captive and field)

| | | | | |
|---|---|---|---|---|
| **Diamondback terrapin (C, SW and FW)** | | | | |
| Malaclemys terrapin | 250 | 10.2 | 91 | in seawater | Robinson and Dunson 1976 |
| | 84 | 20.2 | 400 | juveniles, in seawater | Dunson 1985 |
| | 7.5 | 1.84 | 210 | juveniles, in seawater | |
| | 7.0 | 4.60 | 550 | juveniles, in freshwater | |
| **Desert tortoise (H, X)** | | | | | |
| Gopherus agassizii | 521 | 6.56 | 34 | fed late-spring diet | Nagy and Medica 1986 |

In field

| | | | | | |
|---|---|---|---|---|---|
| **Anole (C, H)** | | | | | |
| Anolis auratus | 1.3 | 0.50 | 630 | adults, rainy season | Nagy et al. (unpubl.) |
| | 0.4 | 0.32 | 950 | juveniles, rainy season | |
| **Anole (C, H)** | | | | | |
| Anolis limifrons | 1.4 | 0.27 | 620 | adults, dry season | Nagy et al. (unpubl.) |
| | 1.6 | 1.03 | 1120 | adults, winter | |
| | 0.4 | 0.25 | 740 | juveniles, winter | |
| | 1.3 | 0.47 | 590 | adults, dry winter | |
| **Side-blotched lizard (C, X)** | | | | | |
| Uta stansburiana | 3.3 | 0.15 | 97 | spring | Nagy and Medica (unpubl.) |
| | 2.8 | 0.07 | 51 | summer | |
| | 3.0 | 0.05 | 34 | autumn | |
| | 2.5 | 0.03 | 24 | winter | |
| **Sand lizard (C, X)** | | | | | |
| Eremias lineoocellata | 3.27 | 0.16 | 104 | spring | Nagy et al. 1984a |
| **Sand lizard (C, X)** | | | | | |
| Eremias lugubris | 3.83 | 0.29 | 170 | spring | Nagy et al. 1984a |
| **Long-tailed brush lizard (C, X)** | | | | | |
| Urosaurus graciosus | 3.6 | 0.132 | 80 | summer | Congdon et al. 1982 |
| **Tree lizard (C, X and M)** | | | | | |
| Urosaurus ornatus | 3.8 | 0.098 | 57 | summer | Congdon et al. 1982 |

Table A1, continued (Reptiles, field)

| GROUP Species (diet,habitat)[1] | Body mass, g | Water flux rate ml/day | % of predicted[2] | Conditions | Reference |
|---|---|---|---|---|---|
| Orange-throated whiptail (C,X) Cnemidophorus hyperythrus | 4.6 4.1 4.2 3.6 | 0.45 0.41 0.22 0.18 | 230 230 120 110 | males, woodland, autumn females, woodland, autumn males, scrub, autumn females, scrub, autumn | Karasov and Anderson 1984 |
| Bunch grass lizard (C,M) Sceloporus scalaris | 4.5 | 0.13 | 67 | summer | Congdon 1977 |
| Striped plateau lizard (C,M) Sceloporus virgatus | 5.3 5.3 | 0.10 0.19 | 46 87 | adults, spring females, summer | Congdon 1977 |
| Sagebrush lizard (C,X and M) Sceloporus graciosus | 6.0 5.2 3.0 | 0.07 0.08 0.05 | 29 37 35 | adults males, summer adults females, summer juveniles | Congdon and Tinkle 1982 |
| Sand-diving lizard (O,X) Aporosaura anchietae | 5.4 3.1 | 0.09 0.07 | 41 47 | males, summer and winter females, summer and winter | Cooper and Robinson (unpubl.) |
| Striped plateau lizard (C,M) Sceloporus virgatus | 5.5 7.6 5.5 6.8 | 0.21 0.32 0.18 0.32 | 93 110 80 120 | males, spring females, spring males, summer females, summer | Merker and Nagy 1984 |
| Zebra-tailed lizard (C,X) Callisaurus draconoides | 6.3 4.6 | 0.16 0.23 | 64 120 | males, summer and autumn females, summer and autumn | Karasov and Anderson (unpubl.) |
| | 8.6 | 0.15 | 48 | summer | Anderson and Karasov 1981 |

94

Table A1, continued (Reptiles, field)

| | | | | |
|---|---|---|---|---|
| Western fence lizard (C,M) | | | | |
| Sceloporus occidentalis | 10.2 | 0.11 | 31 | summer, parietalectomized | Bickler and Nagy 1980 |
| | 10.6 | 0.22 | 61 | summer, sham-operated | |
| | 12.2 | 0.18 | 45 | summer, intact | |
| | 11.9 | 0.23 | 58 | spring | Nagy and Bennett |
| | 11.5 | 0.19 | 49 | autumn | (unpubl.) |
| Yarrow's spiny lizard (C,M) | | | | | |
| Sceloporus jarrovi | 13.0 | 0.16 | 38 | males, winter active period | Congdon et al. 1979 |
| | 9.7 | 0.08 | 24 | females, winter active period | |
| | 12.4 | 0.06 | 15 | males, winter inactive period | |
| | 11.1 | 0.05 | 13 | females, winter inactive period | |
| | 11.1 | 0.42 | 110 | adults, summer | Congdon 1977 |
| | 23.9 | 0.48 | 73 | males, autumn | |
| | 17.4 | 0.32 | 62 | females, autumn | |
| | 8.5 | 0.28 | 91 | juveniles, autumn | |
| Fringe-toed lizard (O,X) | | | | | |
| Uma scoparia | 7 | 0.17 | 63 | juveniles, summer | Minnich and Shoemaker 1972 |
| | 17 | 0.20 | 39 | adults, summer | |
| Western whiptail (C,X) | | | | | |
| Cnemidophorus tigris | 15.7 | 0.58 | 120 | summer | Anderson and Karasov 1981 |
| Ornate dragon (C,X) | | | | | |
| Amphibolorus ornatus | 19.3 | 0.63 | 110 | summer | Baverstock 1975 |
| Central netted dragon (C,X) | | | | | |
| Amphibolorus nuchalis | 23.9 | 1.1 | 170 | summer | Bradshaw 1978 |
| | 23.9 | 0.6 | 92 | winter | |

Table A1, continued (Reptiles, field)

| GROUP Species (diet,habitat)[1] | Body mass, g | Water flux rate ml/day | % of predicted[2] | Conditions | Reference |
|---|---|---|---|---|---|
| Green lacerta (C,M) *Lacerta viridis* | 25.5 | 3.06 | 450 | late summer | Bradshaw et al. 1987 |
| Desert agama (C,X) *Agama mutabilis* | 12.6 | 0.43 | 105 | females, summer | Bradshaw et al. 1976a |
|  | 25.8 | 1.96 | 280 | males, summer |  |
| Lava lizard (C,H) *Tropidurus albemarlensis* | 43.7 | 0.93 | 92 | males, dry season | Nagy (unpubl.) |
|  | 12.7 | 0.30 | 73 | females, dry season |  |
| Desert iguana (H,X) *Dipsosaurus dorsalis* | 47 | 0.85 | 80 | summer | Minnich and Shoemaker 1972 |
|  | 50* | 1.5 | 130 | summer, natural vegetation | Minnich and Shoemaker 1970 |
|  | 50* | 2.4 | 210 | summer, irrigated vegetation |  |
|  | 3.5 | 0.11 | 68 | hatchlings, spring | Mautz and Nagy (unpubl.) |
|  | 50.3 | 0.80 | 71 | adults, spring |  |
|  | 98.1 | 2.32 | 130 | adults, summer |  |
| Chuckwalla (H,X) *Sauromalus obesus* | 150 | 5.43 | 220 | mid-spring | Nagy 1972 |
|  | 140 | 2.39 | 100 | late spring |  |
|  | 120 | 0.12 | 6 | autumn, estivating |  |
|  | 100 | 0.05 | 3 | winter, hibernating |  |
| Green iguana (H,H) *Iguana iguana* | 18.9 | 2.76 | 500 | hatchlings | Nagy et al. (unpubl.) |
|  | 947 | 37.9 | 400 | adults |  |

Table A1, continued (Reptiles, field)

| | | | | |
|---|---|---|---|---|
| Desert tortoise (H,X) | | | | |
| Gopherus agassizii | 950 | 11.4 | 120 | spring |
| | 950 | 7.6 | 80 | summer |
| | 950 | 11.4 | 120 | autumn |
| | 950 | 0.95 | 10 | winter |
| | | | | | Nagy and Medica 1986 |
| | 613 | 1.66 | 24 | summer, dry period |
| | 613 | 2.45 | 36 | summer, after rain | Minnich 1976, 1977 |
| Insular chuckwalla (H,X) | | | | |
| Sauromalus hispidus | 970 | 3.40 | 35 | mid-spring |
| | 900 | 6.30 | 69 | late-spring |
| | 860 | 6.88 | 78 | mid-summer |
| | 840 | 6.30 | 73 | late-summer | Smits 1985 |
| Sand goanna (C,X and M) | | | | |
| Varanus gouldii | 1004 | 15.9 | 160 | spring, temperate |
| | 1004 | 22.1 | 220 | summer, temperate |
| | 1004 | 9.8 | 99 | autumn, temperate |
| | 1004 | 5.5 | 56 | winter, temperate |
| | 1004 | 23.6 | 240 | summer, semiarid | Green 1972 |
| Marine iguana (H,H and SW) | | | | |
| Amblyrhynchus cristatus | 69 | 2.15 | 150 | hatchlings |
| | 1610 | 77.6 | 560 | adults | Shoemaker and Nagy 1984 |
| Gopher tortoise (H,M) | | | | |
| Gopherus polyphemus | 3180 | 98.6 | 430 | summer | Minnich and Ziegler 1977 |
| Perentie (C,X) | | | | |
| Varanus giganteus | 5570 | 32.1 | 94 | autumn |
| | 5570 | 167 | 490 | summer | Green et al. 1986 |

97

Table A1, continued (Reptiles, field and Amphibians, captive)

| GROUP Species (diet,habitat)[1] | Body mass, g | Water flux rate ml/day | % of predicted[2] | Conditions | Reference |
|---|---|---|---|---|---|
| Saltwater crocodile (C,SW and FW) |  |  |  |  |  |
| Crocodylus porosus | 220 | 31.0 | 950 | hatchlings, hyperosmotic estuary | Grigg et al. 1986 |
|  | 190 | 55.3 | 1900 | hatchlings, hyposmotic estuary |  |
|  | 9510 | 231 | 460 | juveniles and subadults, hyperosmotic estuary |  |
|  | 1400 | 124 | 990 | juveniles and subadults, hyposmotic estuary |  |

AMPHIBIANS

In captivity

| Axolotl (FW) |  |  |  |  |  |
|---|---|---|---|---|---|
| Ambystoma mexicanum | 0.23 | 8.94 |  | 10 week old, in water | Dunson et al. 1971 |
| African clawed frog (FW) |  |  |  |  |  |
| Xenopus laevis | 0.08 | 6.4 |  | metamorphosing tadpoles, in water | Schultheiss et al. 1972 |
|  | 0.30 | 21.0 |  | metamorphosing tadpoles, in water |  |
|  | 0.75 | 47.0 |  | metamorphosing tadpoles, in water |  |
|  | 0.40 | 20.0 |  | newly metamorphosed |  |
|  | 0.82 | 54.5 |  | prolactin treated tadpoles |  |
|  | 0.67 | 45.0 |  | hypophysectomized tadpoles |  |
|  | 0.40 | 20.3 |  | hypophysectomized, metamorphosed |  |
|  | 0.41 | 22.3 |  | hypophysectomized, with prolactin |  |
|  | 0.60 | 38.2 |  | hypophysectomized, with ACTH |  |
| Red-spotted newt (FW and M) |  |  |  |  |  |
| Notophthalmus viridescens | 1.54 | 6.64 |  | terrestrial eft form, in water | Robinson and Ewig 1983 |
|  | 2.30 | 6.92 |  | aquatic breeding newt form, in water |  |
|  | 1.31 | 4.17 |  | prolactin-transformed "artificial" newt, |

Table A1, continued (Amphibians, captive and field, and Fishes, captive)

| | | | in water | |
|---|---|---|---|---|
| Leopard frog (FW) | | | | |
| Rana pipiens | 4.13 | 234 | tadpoles, in water | Adams and Peterle 1975 |
| | | | | |
| European fire salamander (M) | | | | |
| Salamandra salamandra | 75 | 3.61 | xeric population, dry soil | Degani 1982 |
| | 75 | 4.09 | xeric population, intermediate soil | |
| | 75 | 7.69 | xeric population, moist soil | |
| | 29 | 2.70 | mesic population, dry soil | |
| | 29 | 2.12 | mesic population, intermediate soil | |
| | 29 | 3.77 | mesic population, moist soil | |

In field

| | | | | |
|---|---|---|---|---|
| European fire salamander (M) | | | | |
| Salamandra salamandra | 69.5 | 29.9 | semi-arid habitat, summer | Degani 1985 |
| | 69.5 | 71.6 | semi-arid habitat, winter | |
| | 30 | 14.4 | moist habitat, summer | |
| | 30 | 27.3 | moist habitat, winter | |

FISHES

In captivity

| | | | | |
|---|---|---|---|---|
| Killifish (FW and SW) | | | | |
| Fundulus kansae | 1.5* | 54.5 | 440 | distilled water, 20°C | Potts and Fleming 1970 |
| | 1.5* | 54.8 | 440 | tap water, 20°C | |
| | 1.5* | 25.1 | 200 | FW, 10 mM Ca, 25 mM Mg, 20°C | |
| | 1.5* | 35.1 | 280 | 33% SW, 20°C | |
| | 1.5* | 29.2 | 230 | 67% SW, 20°C | |
| | 1.5* | 23.5 | 190 | 100% SW, 20°C | |
| | 1.5* | 21.9 | 180 | 150% SW, 20°C | |
| | 1.5* | 20.0 | 160 | 200% SW, 20°C | |
| | 1.5* | 30.5 | 240 | hypophysectomized, FW, 20°C | |
| | 1.5* | 21.1 | 170 | hypophysectomized, SW, 20°C | |
| | 1.5* | 18.4 | 150 | hypophysectomized, 200% SW, 20°C | |
| | 1.5* | 36.5 | 290 | hypophysectomized, prolactin, FW, 20°C | |

Table A1, continued (Fishes, captive)

| GROUP Species (diet,habitat)[1] | Body mass, g | Water flux rate ml/day | % of predicted[2] | Conditions | Reference |
|---|---|---|---|---|---|
| Black-spotted topminnow (FW and SW) |  |  |  |  |  |
| Fundulus olivaceous | 1.5* | 23.9 | 190 | FW, 20°C | Duff and Fleming 1972b |
|  | 1.5* | 16.8 | 134 | 33% SW, 20°C |  |
|  | 1.5* | 13.7 | 110 | 50% SW, 20°C |  |
| Minnow (FW) |  |  |  |  |  |
| Phoxinus phoxinus | 1.6* | 21.5 | 160 | FW, 10°C | Evans 1969b |
|  | 1.6* | 22.2 | 170 | FW, 10°C |  |
|  | 1.6* | 39.3 | 300 | FW, 20°C |  |
| Cyprinodont (FW and SW) |  |  |  |  |  |
| Aphanius dispar | 1.8* | 21.3 | 140 | FW, 20°C | Lotan 1969 |
|  | 1.8* | 28.0 | 190 | SW, 20°C |  |
|  | 1.8* | 21.5 | 150 | 200% SW, 20°C |  |
| Miller's thumb (FW) |  |  |  |  |  |
| Cottus morio | 3.2* | 17.1 | 68 | FW, 10°C | Evans 1969b |
|  | 3.5 | 15.8 | 58 | 75% SW, 10°C | Foster 1969 |
|  | 3.5 | 26.8 | 98 | 10% SW, 10°C |  |
|  | 3.5 | 16.1 | 59 | 36% SW (isosmotic), 10°C |  |
| Ozark studfish (FW and SW) |  |  |  |  |  |
| Fundulus catenatus | 4.0* | 69.8 | 230 | FW, 20°C | Duff and Fleming 1972a |
|  | 4.0* | 86.4 | 280 | 40-65% SW, 20°C |  |
| Three-spined stickleback (FW) |  |  |  |  |  |
| Gasterosteus aculeatus | 4.0* | 12.3 | 40 | FW, 10°C | Evans 1969b |
| Carp (FW) |  |  |  |  |  |
| Cyprinus carpio | 4.4 | 98.2 | 290 | FW, 22-25°C | Adams and Peterle 1975 |

Table A1, continued (Fishes, captive)

| | | | | | |
|---|---|---|---|---|---|
| Whitefish (FW) Coregonus fera | 5.0 | 22.7 | 60 | FW, 16°C | Streit 1980 |
| Prickleback (SW) Lumpenus lampretaeformis | 6.5* | 2.1 | 4 | SW, 10°C | Evans 1969b |
| Gunnelfish (SW and FW) Pholis gunnellus | 6.5*<br>6.5* | 15.2<br>12.9 | 32<br>27 | SW, 10°C<br>20% SW, 10°C | Evans 1969a |
| Bullhead (SW) Cottus bubalis | 7.0<br>7.0<br>7.0 | 21.9<br>17.7<br>23.1 | 43<br>34<br>45 | SW, 10°C<br>10% SW, 10°C<br>36% SW (isosmotic), 10°C | Foster 1969 |
| Bluegill (FW) Lepomis macrochirus | 1.8 | 21.6 | 150 | FW, 22-25°C | Adams and Peterle 1975 |
| | 7.2 | 634 | 1200 | in enclosures in contaminated FW marsh, autumn and winter | Adams et al. 1976 |
| Tilapia (FW and SW) Tilapia mossambica | 10*<br>10*<br>10* | 197<br>144<br>72 | 280<br>200<br>101 | FW, 25°C<br>SW, 25°C<br>200% SW, 25°C | Potts et al. 1967 |
| Gold-sinny wrasse (SW) Ctenolabrus rupestris | 15* | 30.5 | 29 | SW, 10°C | Evans 1969b |
| Roach (FW) Rutilus rutilus | 17* | 235 | 200 | FW, 10°C | Evans 1969b |
| Black goby (SW) Gobius niger | 17* | 15.3 | 13 | SW, 10°C | Evans 1969b |
| Flounder (SW) Platichthys platessa | 20* | 55.2 | 41 | SW, 10°C | Evans 1969b |

Table A1, continued (Fishes, captive)

| GROUP Species (diet, habitat)[1] | Body mass, g | Water flux rate ml/day | % of predicted[2] | Conditions | Reference |
|---|---|---|---|---|---|
| Blenny (SW and FW) Xiphister atropurpureus | 23.5* 23.5* | 58.3 60.9 | 37 39 | SW, 13°C 10% SW, 13°C | Evans 1967 |
| Rainbow trout (FW) Salmo gairdneri | 26* | 160 | 93 | FW, 10°C | Evans 1969b |
| Lesser weaver (SW) Trachinus vipera | 26* | 69.4 | 40 | SW, 10°C | Evans 1969b |
| Salmon (FW and SW) Salmo salar | 33* 33* | 300 203 | 140 95 | FW, 10°C SW, 10°C | Potts et al. 1970 |
| Fresh-water ray (FW) Potamotrygon sp. | 52 | 839 | 260 | FW, 24°C, fasting | Carrier and Evans 1973 |
| Pacific hagfish (SW) Eptatretus stoutii | 54 | 3700 | 1100 | SW, 12°C | Rudy and Wagner 1970 |
| Viviparous blenny (SW) Zoarces viviparous | 66* | 110 | 27 | SW, 10°C | Evans 1969b |
| Brown trout (FW) Salmo trutta | 1.4* 82* | 15.3 495 | 130 100 | FW, 10°C FW, 10°C | Evans 1969b |
| Sea Perch (SW) Serranus cabrilla and S. scriba | 92 40* 40* 40* | 294 198 160 126 | 54 77 63 49 | SW, 16°C SW, 20°C SW, 15°C SW, 10°C | Motais et al. 1969 Isaia 1972 |

102

Table A1, continued (Fishes, captive)

| | | | | | |
|---|---|---|---|---|---|
| Yellow eel (FW and SW) | | | | | |
| Anguilla anguilla | 80* | 118 | 24 | FW, 10°C | Evans 1969b |
| | 1.2 | 2.7 | 27 | FW, 10°C | |
| | | | | | |
| Eel (FW and SW) | | | | | |
| Anguilla anguilla | 129 | 845 | 110 | FW, 19°C | Motais et al. 1969 |
| | 116 | 532 | 78 | SW, 19°C | |
| | 60 | 546 | 150 | SW, SW adapted, 25°C | Motais and Isaia 1972 |
| | 60 | 353 | 95 | SW, SW adapted, 15°C | |
| | 60 | 141 | 38 | SW, SW adapted, 5°C | |
| | 60 | 399 | 110 | FW, FW adapted, 25°C | |
| | 60 | 272 | 73 | FW, FW adapted, 15°C | |
| | 60 | 131 | 35 | FW, FW adapted, 5°C | |
| | 50* | 51.5 | 16 | SW, 13°C | Evans 1969b |
| | 50* | 77.5 | 25 | FW, 13°C | |
| | 50* | 72 | 23 | SW, 13°C | Evans 1969b |
| | 50* | 72 | 23 | FW, 13°C | |
| | | | | | |
| Staghorn sculpin (SW) | | | | | |
| Leptocottus armatus | 141 | 787 | 97 | SW, 12°C | Rudy and Wagner 1970 |
| | | | | | |
| Goldfish (FW) | | | | | |
| Carassius auratus | 5.6* | 43.2 | 103 | FW, 10°C | Evans 1969b |
| | 5.6* | 92.3 | 220 | FW, 20°C | |
| | 5.6* | 35.4 | 84 | FW, 10°C | |
| | 2.5 | 73.1 | 370 | FW, 20°C | Loretz 1979 |
| | 68* | 685 | 160 | FW, 20°C | Lahlou and Sawyer 1969a |
| | 68* | 654 | 160 | FW, sham operated 20°C | |
| | 68* | 604 | 150 | FW, hypophysectomized, 20°C | |
| | 100* | 1090 | 180 | FW, 14°C | Lahlou and Giordan 1970 |

Table A1, continued (Fishes, captive)

| GROUP Species (diet,habitat)[1] | Body mass, g | Water flux rate mL/day | % of predicted[2] | Conditions | Reference |
|---|---|---|---|---|---|
| Goldfish (cont.) | 150* | 3330 | 390 | FW, 25°C | Isaia 1972 |
| | 150* | 1290 | 150 | FW, 15°C | |
| | 150* | 635 | 74 | FW, 5°C | |
| | 96 | 1120 | 200 | FW, 19°C | Motais et al. 1969 |
| Sea scorpion (SW) Cottus scorpius | 150 | 458 | 53 | SW, 10°C | Foster 1969 |
| | 150 | 393 | 46 | 36% SW (isosmotic), 10°C | |
| Nurse shark (SW) Ginglymostoma cirratum | 150 | 2200 | 260 | SW, 24°C | Carrier and Evans 1972 |
| Plaice (SW and FW) Platichthys flesus | 190* | 492 | 46 | SW, sham operated, 13-22°C | Macfarlane and Maetz 1974 |
| | 190* | 370 | 35 | 33% SW, sham operated, 13-22°C | |
| | 190* | 514 | 48 | FW, acclimated < 1 week, sham operated, 13-22°C | |
| | 190* | 737 | 69 | FW, acclimated > 1 week, sham operated, 13-22°C | |
| | 190* | 447 | 42 | SW, hypophysectomized, 13-22°C | |
| | 190* | 434 | 41 | 33% SW, hypophysectomized, 13-22°C | |
| | 190* | 485 | 45 | FW, acclimated < 1 week, hypophysectomized, 13-22°C | |
| | 190* | 737 | 69 | FW, acclimated > 1 week, hypophysectomized, 13-22°C | |
| | 44* | 80.2 | 29 | SW, 10°C | Evans 1969b |
| | 44* | 155 | 56 | SW, 20°C | |
| | 44* | 98 | 35 | SW, 10°C | |
| | 245 | 1150 | 85 | FW, 16°C | Motais et al. 1969 |
| | 220 | 649 | 53 | SW, 16°C | |

Table A1, continued (Fishes, captive)

| | | | | |
|---|---|---|---|---|
| Oysterfish (SW and FW)<br>Opsanus tau | 260*<br>260* | 655<br>328 | 46<br>23 | 5% SW, 20°C<br>SW, 20°C | Lahlou and Sawyer 1969b |
| Dogfish (SW)<br>Scyliorhinus canicula | 300* | 7910 | 490 | SW, 16°C | Payan and Maetz 1971 |
| Electric ray (SW)<br>Torpedo marmorata | 300* | 4890 | 300 | SW, 16°C | Payan and Maetz 1971 |
| Cod (SW)<br>Gadus callarias | 625* | 1380 | 43 | SW, 12°C | Fletcher 1978 |
| Skate (SW)<br>Raja erinacea and<br>R. radiata | 850*<br>850*<br>850*<br>850* | 10700<br>10700<br>10700<br>12400 | 250<br>250<br>250<br>290 | SW, 13°C<br>SW, hypophysectomized, 13°C<br>50% SW, 13°C<br>50% SW, hypophysectomized, 13°C | Payan et al. 1973 |
| Skate (SW)<br>Raja montagu | 500* | 14000 | 540 | SW, 16°C | Payan and Maetz 1971 |
| Dogfish (SW)<br>Poroderma africanum | 2000*<br>2000*<br>2000*<br>2000*<br>2000* | 30600<br>31000<br>34900<br>29200<br>29900 | 330<br>330<br>380<br>310<br>320 | 117% SW, 13°C<br>109% SW, 13°C<br>SW, 13°C<br>91% SW, 13°C<br>81% SW, 13°C | Haywood 1974 |
| White sturgeon (FW and SW)<br>Acipenser transmontanus | 3500<br>3500 | 9000<br>25300 | 58<br>160 | SW, 15°C<br>FW, 19°C | Potts and Rudy 1972 |
| Green sturgeon (FW and SW)<br>Acipenser medirostris | 7700*<br>7700* | 26300<br>53900 | 82<br>170 | SW, 15°C<br>FW, 19°C | Potts and Rudy 1972 |

Table A1, continued (Arthropods, water-breathing, captive)

| GROUP Species (diet,habitat)[1] | Body mass, g | Water flux rate ml/day | % of predicted[2] | Conditions | Reference |
|---|---|---|---|---|---|
| ARTHROPODS: WATER-BREATHING | | | | | |
| In captivity | | | | | |
| Branchiopod (FW) | | | | | |
| Daphnia magna | 0.002* | 0.864 | 96 | FW | Krogh 1939 |
| Caddis-fly larva (SW) | | | | | |
| Philanisus plebeius | 0.0065 | 0.056 | 14 | SW, intact, 20°C | Leader 1972 |
| | 0.0065 | 0.029 | 7 | SW, mouth sealed, 20°C | |
| Crab (SW) | | | | | |
| Pinnixa occidentalis | 0.0132 | 1.20 | 180 | SW, 13°C | Roesijadi 1978 |
| | 0.0184 | 1.06 | 130 | SW, 13°C | |
| | 0.0490 | 2.21 | 130 | SW, 13°C | |
| | 0.0765 | 3.95 | 160 | SW, 13°C | |
| Brine shrimp (SW) | | | | | |
| Artemia salina | 0.03* | 0.09 | 7 | SW | Krogh 1939 |
| | 0.03* | 0.021 | 2 | SW, 20-25°C | Smith 1969 |
| Amphipod (FW) | | | | | |
| Gammarus pulex | 0.04* | 2.72 | 180 | FW, 20°C | Lockwood 1961 |
| Amphipod (SW and FW) | | | | | |
| Gammarus duebeni | 0.04* | 2.27 | 150 | FW, 20°C | Lockwood 1961 |
| | 0.075 | 9.29 | 390 | 150% SW, 18°C | Lockwood et al. 1973 |
| | 0.075 | 7.48 | 310 | 100% SW, 18°C | |
| | 0.075 | 4.23 | 180 | 50% SW, 18°C | |
| | 0.075 | 3.39 | 140 | 2% SW, 18°C | |

Table A1, continued (Arthropods, water-breathing, captive)

| | | | | | |
|---|---|---|---|---|---|
| Amphipod (cont.) | 0.075 | 8.01 | 330 | 2% SW, just moulted | Lockwood and Inman 1973 |
| | 0.075 | 14.4 | 600 | 100% SW, just moulted | |
| Isopod (SW) | | | | | |
| Idotea linearis | 0.12 | 29.9 | 880 | SW, intermoult control | Lockwood and Inman 1973 |
| | 0.12 | 51.9 | 1500 | SW, just moulted | |
| | 0.12 | 25.7 | 750 | SW, 1 week after moult | |
| Isopod (FW) | | | | | |
| Asellus aquaticus | 0.39* | 8.93 | 108 | FW, 18°C | Lockwood 1959 |
| Shrimp (SW) | | | | | |
| Leander squilla | 0.5 | 7.75 | 78 | SW, 13°C | Subramanian 1975 |
| Crab (SW and FW) | | | | | |
| Rhithropanopeus harrisi | 1.0 | 13.8 | 83 | 1% SW, 18-20°C | Smith 1967 |
| | 1.0 | 15.1 | 91 | 5% SW, 18-20°C | |
| | 1.0 | 20.1 | 120 | 50% SW, 18-20°C | |
| | 1.0 | 24.5 | 150 | 95% SW, 18-20°C | |
| | 1.0* | 24.3 | 150 | 75% SW, 20°C | Capen 1972 |
| | 1.0* | 13.7 | 83 | 10% SW, 20°C | |
| Shrimp (SW) | | | | | |
| Crangon vulgaris | 1.0* | 28.3 | 170 | 100% SW | Che Mat and Potts 1985 |
| | 1.0* | 25.9 | 160 | 75% SW | |
| | 1.0* | 25.7 | 160 | 50% SW | |
| | 1.0* | 21.4 | 130 | 25% SW | |
| Prawn (SW and FW) | | | | | |
| Palaemonetes varians | 2.5* | 54.9 | 170 | 5% SW | Parry 1955 |
| | 2.5* | 39.9 | 120 | 70% SW | |
| | 2.5* | 67.7 | 210 | 120% SW | |
| | 2.5* | 26.9 | 82 | 120% SW, 10°C | Rudy 1967 |
| | 2.5* | 26.9 | 82 | 70% SW, 10°C | |
| | 2.5* | 23.1 | 71 | 10% SW, 10°C | |

Table A1, continued (Arthropods, water-breathing, captive)

| GROUP Species (diet,habitat)[1] | Body mass, g | Water flux rate ml/day | % of predicted[2] | Conditions | Reference |
|---|---|---|---|---|---|
| **Fiddler crab (SW)** *Uca rapax* | 2.6<br>2.6<br>2.6 | 9.30<br>9.39<br>9.17 | 28<br>28<br>27 | 100% SW, 24°C<br>50% SW, 24°C<br>3% SW, 24°C | Hannan and Evans 1973 |
| **Fiddler crab (SW)** *Uca pugilator* | 3.0<br>3.0<br>3.0<br>3.0<br>3.0<br>3.0 | 16.7<br>18.1<br>17.3<br>8.0<br>10.5<br>10.1 | 45<br>48<br>46<br>21<br>28<br>27 | 100% SW, 24°C<br>50% SW, 24°C<br>3% SW, 24°C<br>100% SW, 14°C<br>50% SW, 14°C<br>3% SW, 14°C | Hannan and Evans 1973 |
| **Crayfish (FW)** *Procambarus blandingi* | 4.18 | 42.2 | 88 | FW, 22-25°C | Adams and Peterle 1975 |
| **Pink shrimp (SW)** *Penaeus duorarum* | 5.8 | 74.5 | 120 | 100% SW, 24°C | Hannan and Evans 1973 |
| **Horseshoe crab (SW)** *Limulus polyphemus* | 5.9<br>5.9<br>5.9<br>5.9 | 201<br>138<br>158<br>77.4 | 320<br>220<br>250<br>120 | 100% SW, 24°C<br>50% SW, 24°C<br>20% SW, 24°C<br>100% SW, 14°C | Hannan and Evans 1973 |
| **Spider crab (SW)** *Hyas araneus* | 8.0 | 339 | 430 | SW, 13°C | Subramanian 1975 |
| **Purple shore crab (SW)** *Hemigrapsus nudus* | 10*<br>10*<br>10*<br>10* | 239<br>158<br>208<br>128 | 260<br>170<br>230<br>140 | 95% SW, 20°C<br>95% SW, 10°C<br>60% SW, 20°C<br>10% SW, 10°C | Smith and Rudy 1972 |

Table A1, continued (Arthropods, water-breathing, captive)

| | | | | | |
|---|---|---|---|---|---|
| **Arid-zone crab (X and FW)** | | | | | |
| Holthuisana transversa | 10.4 | 29.9 | 32 | FW, adult, intermolt, 25°C | Greenaway (unpubl.) |
| | 9.85 | 53.2 | 58 | FW, adult, late premolt, 25°C | |
| | 5.66 | 51.9 | 86 | FW, adult, early postmolt, 25°C | |
| | 0.006 | 0.25 | 68 | FW, juveniles, 25°C | |
| | 12.2 | 0.57 | 0.5 | in air, 98% rh, 25°C | |
| | 11.5 | 1.52 | 1.5 | in simulated burrows, 25-30°C | |
| **Crab (SW)** | | | | | |
| Macropipus depurator | 20 | 744 | 480 | SW, 13°C | Subramanian 1975 |
| **Crayfish (FW)** | | | | | |
| Astacus fluviatilis | 28 | 94.1 | 47 | FW, 10°C | Rudy 1967 |
| **Crab (SW)** | | | | | |
| Macropipus depurator | 10 | 44.7 | 49 | FW, 13°C | Subramanian 1975 |
| | 30.7 | 1230 | 580 | SW, 10°C | Rudy 1967 |
| **Crab (SW and FW)** | | | | | |
| Carcinus maenas | 40 | 740 | 290 | SW, 13°C | Subramanian 1975 |
| | 39.5 | 524 | 200 | 100% SW, 10°C | Rudy 1967 |
| | 41.1 | 539 | 200 | 70% SW, 10°C | |
| | 41.5 | 502 | 190 | 40% SW, 10°C | |
| | 10 | 396 | 430 | 94% SW, 18°C | Smith 1970a |
| | 10 | 459 | 500 | 75% SW, 18°C | |
| | 10 | 348 | 380 | 50% SW, 18°C | |
| | 10 | 296 | 320 | 30% SW, 18°C | |
| **Crayfish (FW)** | | | | | |
| Astacus leptodactylus | 70 | 311 | 79 | FW, 15°C, intact | Ehrenfeld and Isaia 1974 |
| | 70 | 324 | 83 | FW, 15°C, eyestalk ligated | |

Table A1, continued (Arthropods, water-breathing, captive and field and air-breathing, captive)

| GROUP Species (diet, habitat)[1] | Body mass, g | Water flux rate ml/day | % of predicted[2] | Conditions | Reference |
|---|---|---|---|---|---|
| Blue crab (SW) Callinectes sapidus | 176 209 | 1320 948 | 170 107 | 100% SW 50% SW | Robinson 1982 |
| Crab (SW) Cancer productus | 215 | 8240 | 910 | SW, starved, 12°C | Tucker and Harrison 1974 |
| In field |  |  |  |  |  |
| Arid-zone crab (X and FW) Holthuisana transversa | 10.0 17.7 | 2.90 2.60 | 3 2 | winter, in burrows, moist soil winter, in burrows, dry soil | Greenaway (unpubl.) |
| Crayfish (FW) Procambarus blandingi | 28.3 | 227 | 110 | in contaminated marsh | Adams et al. 1976 |
| ARTHROPODS: AIR-BREATHING |  |  |  |  |  |
| In captivity |  |  |  |  |  |
| House dust mite (M) Dermatophagoides pteronyssinus | 5.8·10⁻⁶ 3.5·10⁻⁶ 5.8·10⁻⁶ 3.5·10⁻⁶ | 3.8·10⁻⁵ 2.7·10⁻⁵ 3.4·10⁻⁵ 2.5·10⁻⁵ | 230 230 210 220 | females, 40% rh, 25°C males, 40% rh, 25°C females, 75% rh, 25°C males, 75% rh, 25°C | Arlian 1975 |
| House dust mite (M) Dermatophagoides farinae | 1.3·10⁻⁵ 1.3·10⁻⁵ 1.3·10⁻⁵ | 4.1·10⁻⁵ 1.9·10⁻⁵ 3.7·10⁻⁵ | 140 66 130 | 0% rh, 25°C 22.5% rh, 25°C 52.5% rh, 25°C | Arlian and Wharton 1974 |

Table A1, continued (Arthropods, air-breathing, captive)

| Species | Values | Conditions | Reference |
|---|---|---|---|
| House dust mite (cont.) | 3.5·10⁻⁶  1.0·10⁻⁷  0.9<br>3.5·10⁻⁶  2.0·10⁻⁷  1.7<br>3.5·10⁻⁶  2.1·10⁻⁶  18.0 | 0% rh, 25°C<br>75% rh, 25°C<br>Protonymphs, 75% rh, 25°C | Ellingsen 1975 |
| Book louse (M)<br>Liposcelis bostrychophilus | 2.8·10⁻⁵  1.3·10⁻⁵  26 | 85% rh, 28°C | Devine 1977 |
| Mite (larvae) (M)<br>Dermacentor variabilis | 4.2·10⁻⁵  5.5·10⁻⁶  8<br>4.2·10⁻⁵  5.1·10⁻⁶  8<br>4.2·10⁻⁵  5.6·10⁻⁶  9<br>4.2·10⁻⁵  4.9·10⁻⁶  7 | 0% rh, 25°C<br>53% rh, 25°C<br>85% rh, 25°C<br>92% rh, 25°C | Knulle and Devine 1972 |
| Spiny rat mite (M)<br>Laelaps echidnina | 1.9·10⁻⁴*  1.3·10⁻⁴  69 | 92.5% rh, 25°C, starved | Wharton and Devine 1968 |
| | 1.9·10⁻⁴*  1.0·10⁻⁴  53 | 0% rh, 25°C, starved | Devine and Wharton 1973 |
| Fruit fly (M)<br>Drosophila pseudoobscura | 8.65·10⁻⁴  1.21·10⁻³  220<br>8.65·10⁻⁴  0.73·10⁻³  140<br>8.65·10⁻⁴  0.66·10⁻³  120<br>7.33·10⁻⁴  1.14·10⁻³  240<br>7.33·10⁻⁴  0.61·10⁻³  130<br>7.33·10⁻⁴  0.31·10⁻³  64 | females, 99% rh, 25°C<br>females, 90% rh, 25°C<br>females, 50% rh, 25°C<br>males, 99% rh, 25°C<br>males, 90% rh, 25°C<br>males, 50% rh, 25°C | Arlian and Eckstrand 1975 |
| Flour beetle (M)<br>Tribolium confusum | 0.0018  0.00020  22<br>0.0018  0.00016  18<br>0.0018  0.00012  13<br>0.0018  0.00007  8 | 99% rh<br>85% rh<br>65% rh<br>22.5% rh | Arlian and Veselica 1979 |
| Rice weevil (M)<br>Sitophilus oryzae | 0.0021  0.00023  23 | 22.5–99% rh | Arlian 1979 |

Table A1, continued (Arthropods, air-breathing, captive)

| GROUP Species (diet, habitat)[1] | Body mass, g | Water flux rate ml/day | % of predicted[2] | Conditions | Reference |
|---|---|---|---|---|---|
| Semiaquatic beetle (M and FW) | | | | | |
| Peltodytes muticus | 0.0051 | 0.0056 | 300 | in FW, 10°C | Arlian and Veselica 1979 |
| | 0.0051 | 0.0103 | 550 | in FW, 23°C | Arlian and Staiger 1979 |
| | 0.0051 | 0.0122 | 660 | in FW, 30°C | |
| | 0.0051 | 0.0053 | 280 | in SW, 23°C | |
| | 0.0051 | 0.0059 | 320 | in air, 0% rh, 23°C | |
| | 0.0051 | 0.0116 | 620 | in air, 0% rh, 30°C | |
| | 0.0051 | 0.0077 | 410 | in air, 75% rh, 30°C | |
| | 0.0051 | 0.0089 | 480 | in air, 75% rh, 30°C | |
| | 0.0051 | 0.0072 | 390 | in air, 95% rh, 30°C | |
| Water bug (FW) | | | | | |
| Sigara lateralis | 0.0073 | 0.122 | 5100 | FW, adult females, 16°C | Streit 1980 |
| Sigara striata | 0.0007 | 0.051 | 1100 | FW, juveniles, 16°C | |
| Mosquito (larvae) (SW) | | | | | |
| Opifex fuscus | 0.0076 | 0.039 | 1600 | SW, fourth instar | Nicolson and Leader 1974 |
| Ant (M) | | | | | |
| Formica exsectoides | 0.0086 | 0.0022 | 82 | spring, 24.5°C | Sigal and Arlian 1982 |
| | 0.0086 | 0.0077 | 290 | summer, 24.5°C | |
| | 0.0086 | 0.0194 | 720 | early fall, 24.5°C | |
| | 0.0086 | 0.0230 | 860 | late fall, 24.5°C | |
| Mosquito (larvae) (FW) | | | | | |
| Sialis lutaria | 0.065 | 0.084 | 770 | FW, 8°C | Shaw 1955 |
| | 0.065 | 0.42 | 3800 | FW, 20°C | |
| Beetle (larvae) (M) | | | | | |
| Tenebrio molitor | 0.075* | 0.0041 | 34 | 85-100% rh | Marcuzzi and Santoro 1959 |

112

Table A1, continued (Arthropods, air-breathing, captive)

| | | | | |
|---|---|---|---|---|
| Water bug (FW)<br>Corixa dentipes | 0.090 | 0.545 | 4000 | FW, 18°C | Staddon 1966 |
| Isopod (M)<br>Oniscus asellus | 0.10*<br>0.10*<br>0.10* | 0.235<br>0.288<br>0.401 | 1700<br>1900<br>2700 | 100% rh, dry substratum<br>100% rh, moist substratum<br>FW | Mayes and Holdrich 1975 |
| Isopod (H)<br>Ligia oceanica | 0.10* | 0.379 | 2600 | 100% rh, dry substratum | Mayes and Holdrich 1975 |
| Isopod (M)<br>Porcellio scaber | 0.10* | 0.149 | 1000 | 100% rh, dry substratum | Mayes and Holdrich 1975 |
| Isopod (M)<br>Armadillidium vulgare | 0.10* | 0.103 | 700 | 100% rh, dry substratum | Mayes and Holdrich 1975 |
| Brown paper wasp (M)<br>Polistes fuscatus | 0.122 | 0.020 | 120 | 50% rh, 25°C | King (unpubl.) |
| Water bug (FW)<br>Ilyocoris cimicoides | 0.13<br>0.13<br>0.13 | 0.227<br>0.408<br>0.612 | 1300<br>2300<br>3400 | FW, 8°C<br>FW, 18°C<br>FW, 28°C | Staddon 1966 |
| Water bug (FW)<br>Notonecta glauca | 0.136 | 0.362 | 2000 | FW, 18°C | Staddon 1966 |
| Cricket (M)<br>Acheta domesticus | 0.3* | 0.0462 | 140 | 25°C | Van Hook and Deal 1973 |

Table A1, continued (Arthropods, air-breathing, captive and field)

| GROUP Species (diet, habitat)[1] | Body mass, g | Water flux rate mL/day | % of predicted[2] | Conditions | Reference |
|---|---|---|---|---|---|
| Darkling beetle (X) *Eleodes armata* | 0.60 | 0.031 | 60 | simulated summer burrow | Bohm and Hadley 1977 |
| | 0.81 | 0.024 | 38 | simulated winter burrow | |
| | 0.82 | 0.038 | 59 | simulated summer burrow, no food and water | |
| | 1.06 | 0.194 | 250 | simulated summer burrow, food and water ad lib. | |
| | 1.03 | 0.020 | 27 | simulated winter burrow, no food or water | |
| | 0.88 | 0.026 | 39 | simulated winter burrow, food and water ad lib. | |
| | 1.0* | 0.038 | 52 | 40% rh, 30°C | Cooper 1983 |
| Locust (M and X) *Locusta migratoria* | 1.2* | 0.039 | 47 | 25°C | Buscarlet and Proux 1975 Buscarlet et al. 1978 |
| Desert scorpion (X) *Hadrurus arizonensis* | 3.5* | 0.0088 | 5 | starved, 20°C | King and Hadley 1979 |
| | 3.5* | 0.0158 | 9 | starved, 25°C | |
| | 3.5* | 0.0231 | 13 | starved, 30°C | |
| | 3.5* | 0.0308 | 17 | starved, 35°C | |
| | 3.5* | 0.0490 | 28 | fed, steady state, 35°C | |
| In field | | | | | |
| Brown paper wasp (M) *Polistes fuscatus* | 0.161 | 0.067 | 320 | queens | King (unpubl.) |
| | 0.122 | 0.050 | 290 | workers | |
| | 0.156 | 0.052 | 260 | males | |
| | 0.138 | 0.032 | 170 | females | |

Table A1, continued (Arthropods, air-breathing, field)

| | | | | | |
|---|---|---|---|---|---|
| Desert beetle (X) | | | | | |
| Cysteodemus armatus | 0.271 | 0.125 | 420 | spring | Cohen et al. 1981 |
| Desert beetle (X) | | | | | |
| Cryptoglossa verrucosa | 0.60 | 0.0060 | 12 | late summer | Cooper 1985 |
| | 0.65 | 0.0020 | 4 | winter | |
| | 0.69 | 0.0028 | 5 | spring | |
| | 0.69 | 0.0200 | 35 | early summer | |
| | 0.65 | 0.0028 | 5 | winter, confined to burrow | |
| | 0.69 | 0.0060 | 11 | summer, confined to burrow | |
| Darkling beetle (X) | | | | | |
| Eleodes armata | 0.84 | 0.066 | 101 | early summer | Bohm and Hadley 1977 |
| | 0.79 | 0.049 | 78 | late summer | |
| | 0.70 | 0.036 | 63 | summer | |
| | 0.95 | 0.030 | 42 | winter | |
| | 0.80 | 0.027 | 43 | winter | |
| | 0.93 | 0.033 | 47 | late summer | Cooper 1985 |
| | 0.77 | 0.0039 | 6 | early summer | |
| | 0.81 | 0.012 | 19 | mid-winter | |
| | 1.03 | 0.039 | 52 | spring | |
| | 1.01 | 0.051 | 69 | early summer | |
| | 0.81 | 0.007 | 11 | winter, confined to burrow | |
| | 1.03 | 0.019 | 25 | spring, confined to burrow | |
| | 1.01 | 0.035 | 47 | summer, confined to burrow | |
| Desert beetle (X) | | | | | |
| Onymacris unguicularis | 0.62 | 0.031 | 59 | summer | Cooper 1982 |
| Desert scorpion (X) | | | | | |
| Paruroctonus mesaensis | 3.1 | 0.09 | 55 | spring | Yokota 1979 |
| | 2.4 | 0.05 | 37 | fall | |
| Desert scorpion (X) | | | | | |
| Hadrurus arizonensis | 3.5* | 0.094 | 53 | early summer | King and Hadley 1979 |
| | 3.5* | 0.137 | 78 | late summer | |
| | 3.5* | 0.265 | 150 | fall | |

Table A1, continued (Mollusks, captive)

| GROUP Species (diet, habitat)[1] | Body mass, g | Water flux rate ml/day | % of predicted[2] | Conditions | Reference |
|---|---|---|---|---|---|
| **MOLLUSKS** | | | | | |
| *In captivity* | | | | | |
| Limpet (FW) | | | | | |
| <u>Ancylus fluviatilis</u> | 0.008 | 2.69 | 73 | FW, 16°C | Streit 1980 |
| Pond snail (FW) | | | | | |
| <u>Lymnaea exilis</u> | 0.99 | 102 | 150 | FW, 24-26°C | Adams and Peterle 1975 |
| Arca (SW) | | | | | |
| <u>Anadara granosa</u> | 3 | 166 | 120 | SW | Soman and Krishnamoorthy 1973 |
| | 4.25 | 203 | 120 | SW | |
| | 7.5 | 310 | 130 | SW | |
| Pond snail (FW) | | | | | |
| <u>Viviparus malleatus</u> | 24 | 599 | 120 | FW, 24-26°C | Adams and Peterle 1975 |
| Oyster (SW) | | | | | |
| <u>Crassostrea gigas</u> | 35 | 358 | 58 | SW, starved, 12°C | Tucker and Harrison 1974 |
| | 35 | 668 | 105 | SW, fed algae, 12°C | |
| Clam (SW) | | | | | |
| <u>Mya arenaria</u> | 40 | 272 | 40 | SW, starved, 12°C | Tucker and Harrison 1974 |
| | 40 | 1156 | 170 | SW, fed algae, 12°C | |

Table A1, continued (Annelids, captive)

## ANNELIDS

### In captivity

**Polychaete worm (FW and SW)**
*Nereis diversicolor*

| | | | |
|---|---|---|---|
| 0.3* | 44.3 | estuary population, FW, 18-19°C | Smith 1970b |
| 0.3* | 37.5 | as above, 2% SW | |
| 0.3* | 59.6 | as above, 10% SW | |
| 0.3* | 66.0 | as above, 25% SW | |
| 0.3* | 73.5 | as above, 50% SW | |
| 0.3* | 55.6 | as above, 5% SW, 22-23°C | |
| 0.3* | 70.1 | as above, 18% SW | |
| 0.3* | 58.2 | marine population, 25% SW, 18-19°C | |
| 0.3* | 57.4 | as above, 50% SW | |
| 0.3* | 72.0 | as above, 75% SW | |
| 0.3* | 74.7 | as above, 94% SW | |
| 0.53* | 53.2 | SW acclimated, 150% SW, 12°C | Fletcher 1974 |
| 0.53* | 57.1 | as above, 120% SW | |
| 0.53* | 58.5 | as above, SW | |
| 0.53* | 57.8 | as above, 80% SW | |
| 0.53* | 62.2 | as above, 60% SW | |
| 0.53* | 43.6 | 20% SW acclimated, FW, 12°C | |
| 0.53* | 37.7 | as above, 10% SW | |
| 0.53* | 37.7 | as above, 20% SW | |
| 0.53* | 38.1 | as above, 30% SW | |
| 0.53* | 41.3 | as above, 40% SW | |
| 0.53* | 47.8 | 20% SW, 14°C | Fletcher 1974 |
| 0.53* | 44.9 | Calcium-free 20% SW, 12°C | |
| 0.53* | 39.3 | as above after 24 hours | |

**Polychaete worm (FW and SW)**
*Nereis succinea*

| | | | |
|---|---|---|---|
| 0.5 | 97.6 | 50% SW | Smith 1964 |
| 0.5 | 86.8 | 20% SW | |
| 0.5 | 82.8 | 5% SW | |

117

Table A1, continued (Annelids, captive and Single cells)

| GROUP Species (diet, habitat)[1] | Body mass, g | Water flux rate ml/day | % of predicted[2] | Conditions | Reference |
|---|---|---|---|---|---|
| Polychaete worm (FW) Nereis limnicola | 0.5 0.5 0.5 | 60.4 53.9 51.3 | | 50% SW 20% SW FW | Smith 1964 |
| SINGLE CELLS | | | | | |
| Erythrocyte, mammal Homo sapiens | $9.0 \cdot 10^{-8}$* | 1.26 | | isotonic saline | Solomon 1960 |
| | $9.0 \cdot 10^{-8}$* | 1.09 | | isotonic saline, 23°C | Paganelli and Solomon 1957 |
| Smooth muscle fiber, mammal Cavia porcellus | $1.0 \cdot 10^{-5}$* $1.0 \cdot 10^{-5}$* $1.0 \cdot 10^{-5}$* | 0.080 0.016 0.022 | | still, 25°C agitated, 25°C still, 37°C | Elford 1972 |
| Striated muscle fiber Rana temporaria | $2.0 \cdot 10^{-5}$* $2.0 \cdot 10^{-5}$* $2.0 \cdot 10^{-5}$* | 0.012 0.025 0.008 | | still, 20°C agitated, 20°C still, 10°C | Elford 1972 |
| Egg, fish Pleuronectes platessa and Solea vulgaris | 0.005 0.005 0.005 0.005 | 0.110 0.022 0.0022 0.0055 | | inside female, 8°C in SW 1 hour, 8°C in SW 24 hour, 8°C in SW 8 day, 8°C | Potts and Eddy 1973 |

Table A1, continued (Single cells)

| | | | | |
|---|---|---|---|---|
| Egg, amphibian | | | | |
| Ambystoma mexicanum | 0.021* | 2.67 | oocytes, 20°C | Haglund and Lovtrup 1966 |
| | 0.021* | 1.77 | oocytes, 4°C | |
| | 0.021* | 1.31 | unfertilized egg, 24°C | |
| | 0.021* | 0.26 | unfertilized egg, 4°C | |
| Egg, fish | | | | |
| Salmo salar | 0.088 | 0.68 | river water, freshly stripped, 3.5°C | Potts and Rudy 1969 |
| | 0.088 | 0.66 | distilled water, as above | |
| | 0.088 | 0.51 | isotonic saline, as above | |
| | 0.088 | 1.22 | isotonic glucose, as above | |
| | 0.108 | 0.17 | river water, 2 hour old, 3.5°C | |
| | 0.110 | 0.19 | distilled water, as above | |
| | 0.088 | 0.29 | isotonic saline, as above | |
| | 0.113 | 0.12 | isotonic glucose, as above | |

TABLE A2. Regression statistics for allometry of water flux rate in animals. The values shown include the standard error (SE) and the 95% confidence intervals (95% CI) of the intercept (log $a$) and slope ($b$) of the regressions, number of data points ($N$), coefficient of determination ($r^2$), probability value ($P$) for significance of regression from F-statistic, and mean values of $\log_{10} x$ and $\log_{10} y$ ($\overline{\log x}$, $\overline{\log y}$).

| Group | log $a$ | (SE$_{\log a}$) | 95% CI log $a$ | $b$ | (SE$_b$) | 95% CI $b$ | $N$ | $r^2$ | $P$ | $\overline{\log x}$ | $\overline{\log y}$ | 95% CI of predicted log $y$[1] $c$ | $d$ | $e$ |
|---|---|---|---|---|---|---|---|---|---|---|---|---|---|---|
| **EUTHERIAN MAMMALS** | | | | | | | | | | | | | | |
| In captivity | -0.798 | (0.035) | -0.867 -0.729 | .946 | (0.009) | .930 .963 | 562 | .957 | <.001 | 3.827 | 2.823 | 0.604 | 1.002 | 7.69x10$^{-4}$ |
| In field | -0.487 | (0.071) | -0.628 -0.346 | .818 | (0.027) | .765 .870 | 115 | .894 | <.001 | 2.380 | 1.459 | 0.694 | 1.009 | 5.74x10$^{-3}$ |
| Herbivores | -0.150 | (0.094) | -0.338 0.037 | .795* | (0.024) | .747 .844 | 28 | .906 | <.001 | 3.159 | 2.362 | 1.985 | 0.087 | 5.93x10$^{-4}$ |
| Omnivores | -0.471 | (0.061) | -0.591 -0.350 | .795* | (0.024) | .747 .844 | 47 | .906 | <.001 | 1.784 | 0.948 | 1.985 | 0.086 | 5.93x10$^{-4}$ |
| Carnivores & granivores | -0.605 | (0.078) | -0.760 -0.449 | .795* | (0.024) | .747 .844 | 39 | .906 | <.001 | 2.598 | 1.462 | 1.985 | 0.086 | 5.93x10$^{-4}$ |
| Desert eutherians | -0.838 | (0.082) | -1.002 -0.674 | .954 | (0.039) | .877 1.032 | 67 | .903 | <.001 | 1.968 | 1.040 | 0.482 | 1.015 | 2.59x10$^{-2}$ |

TABLE A2, continued

| | | | | | | | | | | |
|---|---|---|---|---|---|---|---|---|---|---|
| MARSUPIAL MAMMALS | | | | | | | | | | |
| In captivity | -0.262 (0.129) | -0.525 0.001 | .771 (0.042) | .685 .857 | 36 | .907 | <.001 | 2.863 | 1.944 | 0.549 | 1.028 | $2.45 \times 10^{-2}$ |
| In field | 0.396 (0.074) | 0.248 0.544 | .602 (0.024) | .553 .651 | 57 | .917 | <.001 | 2.832 | 2.101 | 0.381 | 1.018 | $1.65 \times 10^{-2}$ |
| Herbivores | -0.058 (0.124) | -0.307 0.190 | .711* (0.033) | .644 .777 | 28 | .907 | <.001 | 3.642 | 2.530 | 2.011 | 0.027 | $1.08 \times 10^{-3}$ |
| Carnivores | 0.270 (0.075) | 0.119 0.421 | .711* (0.033) | .644 .777 | 23 | .907 | <.001 | 2.042 | 1.721 | 2.011 | 0.028 | $1.08 \times 10^{-3}$ |
| BIRDS | | | | | | | | | | |
| In captivity | -0.059 (0.083) | -0.222 0.105 | .694* (0.027) | .642 .747 | 74 | .838 | <.001 | 2.763 | 1.860 | 1.978 | 0.110 | $7.01 \times 10^{-4}$ |
| In field | 0.137 (0.071) | -0.003 0.276 | .694* (0.027) | .642 .747 | 62 | .838 | <.001 | 2.148 | 1.628 | 1.978 | 0.110 | $7.01 \times 10^{-4}$ |
| Passerines | 0.070 (0.090) | -0.111 0.250 | .876* (0.055) | .766 .987 | 26 | .811 | <.001 | 1.301 | 1.210 | 2.000 | 0.082 | $3.04 \times 10^{-3}$ |
| Carnivores | -0.008 (0.228) | -0.481 0.464 | .746 (0.078) | .585 .908 | 24 | .807 | <.001 | 2.725 | 2.026 | 0.845 | 1.042 | $3.65 \times 10^{-2}$ |
| Desert birds | -0.020 (0.122) | -0.264 0.224 | .674* (0.041) | .592 .757 | 18 | .821 | <.001 | 2.371 | 1.579 | 2.000 | 0.103 | $1.69 \times 10^{-3}$ |
| Seabirds | -0.568 (0.380) | -1.373 0.237 | .902 (0.116) | .656 1.148 | 18 | .791 | <.001 | 3.162 | 2.284 | 0.896 | 1.056 | $7.53 \times 10^{-2}$ |

TABLE A2, continued

| Group | log a | (SE_log a) | 95% CI log a | b | (SE_b) | 95% CI b | N | $r^2$ | P | log x | log y | 95% CI of predicted log y[1] c | d | e |
|---|---|---|---|---|---|---|---|---|---|---|---|---|---|---|
| **REPTILES** | | | | | | | | | | | | | | |
| In captivity | -0.691 | (0.117) | -0.922 -0.460 | .726* | (0.047) | .634 .818 | 41 | .648 | <.001 | 1.809 | 0.622 | 1.978 | 0.278 | 2.19x10⁻³ |
| In field | -1.185 | (0.089) | -1.359 -1.011 | .726* | (0.047) | .634 .818 | 93 | .648 | <.001 | 1.498 | -0.097 | 1.978 | 0.274 | 2.19x10⁻³ |
| Desert reptiles | -1.421 | (0.103) | -1.626 -1.217 | .792* | (0.048) | .697 .888 | 55 | .751 | <.001 | 1.663 | -0.104 | 1.987 | 0.232 | 2.31x10⁻³ |
| **AMPHIBIANS** | | | | | | | | | | | | | | |
| In captivity | | | | | | | | | | | | | | |
| Water-breathing larvae | 1.765 | (0.035) | 1.687 1.843 | .980 | (0.068) | .826 1.135 | 11 | .958 | <.001 | -0.312 | 1.459 | 0.205 | 1.091 | 5.68x10⁻¹ |
| **FISHES** | | | | | | | | | | | | | | |
| In captivity[2] | 0.935 | (0.071) | 0.795 1.075 | .919 | (0.039) | .841 .997 | 121 | .821 | <.001 | 1.496 | 2.310 | 0.851 | 1.008 | 8.41x10⁻³ |
| SW osteichthyes in SW | 0.389 | (0.129) | 0.130 0.648 | .990* | (0.065) | .860 1.121 | 14 | .833 | <.001 | 1.607 | 1.980 | 2.012 | 0.086 | 4.19x10⁻³ |
| FW osteichthyes in FW | 1.043 | (0.089) | 0.864 1.223 | .990* | (0.065) | .860 1.121 | 26 | .833 | <.001 | 1.080 | 2.113 | 2.012 | 0.084 | 4.19x10⁻³ |
| SW chondrichthyes in SW | 1.247 | (0.208) | 0.828 1.665 | .990* | (0.065) | .860 1.121 | 11 | .833 | <.001 | 2.927 | 4.145 | 2.012 | 0.088 | 4.19x10⁻³ |

TABLE A2, continued

## ARTHROPODS

### In captivity

| | | | | | | | | | | |
|---|---|---|---|---|---|---|---|---|---|---|
| Water-breathers[2] | 1.219 (0.072) | 1.076 1.362 | .745 (0.061) | .623 .868 | 84 | .642 | <.001 | 0.230 1.390 | 1.284 1.012 | $9.06 \times 10^{-3}$ |
| SW arthropods in SW | 1.405 (0.123) | 1.152 1.659 | .967 (0.094) | .774 1.160 | 28 | .803 | <.001 | -0.190 1.222 | 1.328 1.036 | $2.12 \times 10^{-2}$ |
| FW arthropods in FW | 1.203 (0.063) | 1.066 1.341 | .616 (0.042) | .524 .708 | 14 | .947 | <.001 | 0.203 1.329 | 0.509 1.071 | $3.26 \times 10^{-2}$ |
| Air-breathers | -1.132 (0.112) | -1.355 -0.909 | .697 (0.047) | .605 .789 | 103 | .689 | <.001 | -1.624 -2.264 | 1.675 1.010 | $3.04 \times 10^{-3}$ |
| Aquatic arthropods in water | 0.380 (0.175) | 0.033 0.728 | .881* (0.054) | .774 .987 | 14 | .746 | <.001 | -1.680 -1.100 | 1.989 0.337 | $2.87 \times 10^{-3}$ |
| Mesic arthropods in air | -0.671 (0.191) | -1.052 -0.291 | .881* (0.054) | .774 .987 | 43 | .746 | <.001 | -3.198 -3.489 | 1.989 0.321 | $2.87 \times 10^{-3}$ |
| Desert arthropods in air | -1.635 (0.090) | -1.813 -1.456 | .881* (0.054) | .774 .987 | 39 | .746 | <.001 | 0.059 -1.583 | 1.989 0.322 | $2.87 \times 10^{-3}$ |

## MOLLUSKS

| | | | | | | | | | | |
|---|---|---|---|---|---|---|---|---|---|---|
| In captivity: aquatic | 1.849 (0.080) | 1.665 2.033 | .612 (0.060) | .473 .750 | 10 | .929 | <.001 | 0.755 2.311 | 0.478 1.100 | $8.41 \times 10^{-2}$ |

[1] Equation for calculating the 95% CI of a predicted log $y$ value at any log $x$ value is of the form:

$$95\% \text{ CI}_{\log y} = \log y_p \pm c \, [d + e \, (\log x - \overline{\log x})^2]^{0.5}$$

[2] SW = seawater; FW = freshwater.

* Common slope from ANCOVA.

TABLE A3. Summary of water economy index (WEI) values (ratios of water flux to metabolic rate) in free-living animals. Abbreviations as in Table A1.

| GROUP<br>Species | | Body mass, g | Water flux, ml/day | Energy metabolism, kJ/day | WEI, ml/kJ |
|---|---|---|---|---|---|
| **EUTHERIAN MAMMALS** | | | | | |
| House mouse (O,M) | *Mus musculus* | 13.0 | 3.30 | 39.8 | 0.083* |
| Bat (C,X) | *Macrotus californicus* | 12.6 | 3.0 | 22.2 | 0.135 |
| | | 13.3 | 1.8 | 20.8 | 0.086 |
| Pocket mouse (G,X) | *Perognathus formosus* | 18.0 | 1.31 | 26.8 | 0.049 |
| | | 18.0 | 2.23 | 41.2 | 0.054 |
| Vole (O,M) | *Clethrionomys rutilus* | 13.6 | 3.6 | 59.0 | 0.061 |
| | | 18.3 | 4.9 | 52.8 | 0.093 |
| | | 15.9 | 5.7 | 60.2 | 0.095 |
| | | 16.1 | 5.8 | 58.5 | 0.099 |
| Kangaroo rat (G,X) | *Dipodomys merriami* | 31.7 | 4.2 | 37.3 | 0.113 |
| | | 32.0 | 3.2 | 52.9 | 0.060 |
| | | 38.5 | 3.1 | 63.6 | 0.049 |
| Kangaroo rat (G,X) | *Dipodomys microps* | 51.9 | 6.5 | 34.5 | 0.188 |
| | | 61.0 | 7.6 | 102 | 0.075 |
| | | 55.1 | 10.7 | 137 | 0.078 |
| | | 55.0 | 6.2 | 64.5 | 0.096 |
| Grey mouse (O,M) | *Pseudomys albocinereus* | 32.6 | 7.76 | 62.2 | 0.125 |
| Spiny mouse (O,X) | *Acomys cahirinus* | 38.3 | 5.06 | 51.8 | 0.098* |
| Spiny mouse (O,X) | *Acomys russatus* | 45.0 | 5.65 | 47.6 | 0.119* |
| Bushy-tailed jird (O,X) | *Sekeetamys calurus* | 41.2 | 5.89 | 44.0 | 0.134* |
| Ground squirrel (O,X)[1] | *Ammospermophilus leucurus* | 90.0 | 18.0 | 114 | 0.158* |
| | | 79.9 | 11.9 | 79.3 | 0.150* |
| | | 82.1 | 9.03 | 79.6 | 0.113* |
| | | 96.1 | 8.65 | 79.0 | 0.109* |
| Vole (H,H and M) | *Arvicola terrestris* | 84.4 | 62.9 | 149 | 0.422 |
| | | 93.6 | 78.3 | 88.7 | 0.883 |
| Pocket gopher (H,M) | *Thomomys bottae* | 99.4 | 26 | 127 | 0.205* |
| | | 108 | 27 | 128 | 0.211* |
| | | 104 | 53 | 136 | 0.390* |
| Jackrabbit (H,X)[1] | *Lepus californicus* | 1800 | 82.4 | 1420 | 0.058 |
| | | 1800 | 67.3 | 1180 | 0.057 |
| Three-toed sloth (H,M) | *Bradypus variegatus* | 3830 | 154 | 590 | 0.261* |
| | | 4320 | 125 | 490 | 0.255* |
| | | 4450 | 181 | 739 | 0.245* |

TABLE A3, continued

| GROUP<br>Species | | Body mass, g | Water flux, ml/day | Energy metabolism, kJ/day | WEI, ml/kJ |
|---|---|---|---|---|---|
| **EUTHERIAN MAMMALS (cont.)** | | | | | |
| Howler monkey (H,M)[1] | *Alouatta palliata* | 6500 | 767 | 2308 | 0.332* |
| Fur seal (C,SW) | *Callorhinus ursinus* | 30900 | 5690 | 21800 | 0.261 |
| Deer (H,M) | *Odocoileus hemionus* | 39975 | 4550 | 23400 | 0.194* |
| | | 67100 | 8050 | 40000 | 0.201* |
| Sea lion (C,SW) | *Zalophus californianus* | 82800 | 7620 | 38400 | 0.198 |
| | | 75200 | 7640 | 50200 | 0.152 |
| **MARSUPIAL MAMMALS** | | | | | |
| Dunnart (C,M) | *Sminthopsis crassicaudata* | 6.1 | 3.46 | 29.1 | 0.119 |
| | | 16.6 | 13.4 | 68.8 | 0.195* |
| Brown antechinus (C,M) | *Antechinus stuartii* | 25.7 | 13.9 | 67.3 | 0.207 |
| | | 25.7 | 18.9 | 87.3 | 0.216 |
| Marsupial mouse (C,M) | *Antechinus swainsonii* | 26.3 | 15.0 | 66.2 | 0.227* |
| | | 32.1 | 18.1 | 92.1 | 0.197* |
| | | 47.4 | 23.1 | 126 | 0.183* |
| | | 54.2 | 72.5 | 221 | 0.328* |
| | | 52.5 | 36.4 | 124 | 0.294 |
| | | 72.7 | 50.4 | 177 | 0.285* |
| Sugar glider (O,M) | *Petaurus breviceps* | 112 | 21.7 | 153 | 0.142* |
| | | 135 | 40.6 | 192 | 0.211* |
| Leadbeater's possum (O,M) | *Gymnobelideus leadbeateri* | 118 | 37.9 | 219 | 0.173* |
| | | 135 | 32.6 | 232 | 0.141* |
| Ringtail possum (H,M) | *Pseudocheirus peregrinus* | 278 | 43.6 | 249 | 0.175 |
| | | 717 | 106 | 556 | 0.191* |
| Greater glider (H,M) | *Petauroides volans* | 934 | 80.4 | 690 | 0.117* |
| | | 1042 | 94.8 | 570 | 0.166* |
| Bandicoot (C,M) | *Isoodon obesulus* | 1232 | 109 | 690 | 0.158* |
| Quokka wallaby (H,M) | *Setonix brachyurus* | 1507 | 71.7 | 486 | 0.148* |
| | | 2472 | 116 | 662 | 0.175* |
| Tammar wallaby (H,M) | *Macropus eugenii* | 4560 | 262 | 1230 | 0.213* |
| Pademelon wallaby (H,M) | *Thylogale billardieri* | 5450 | 621 | 1490 | 0.417* |
| Koala (H,M) | *Phascolarctos cinereus* | 7800 | 358 | 2050 | 0.175* |
| | | 10800 | 475 | 2030 | 0.234* |

TABLE A3, continued

| GROUP  Species | | Body mass, g | Water flux, ml/day | Energy metabolism, kJ/day | WEI, ml/kJ |
|---|---|---|---|---|---|
| MARSUPIAL MAMMALS (cont.) | | | | | |
| Grey kangaroo (H,M) | *Macropus giganteus* | 27000 | 1570 | 5610 | 0.280* |
| | | 61900 | 3840 | 11700 | 0.328* |
| BIRDS | | | | | |
| Anna's hummingbird (N,M) | *Calypte anna* | 4.5 | 5.1 | 26.7 | 0.191 |
| Silvereye (O,M) | *Zosterops lateralis* | 9.1 | 5.53 | 35.3 | 0.157 |
| | | 9.7 | 11.6 | 39.2 | 0.296 |
| | | 9.6 | 12.8 | 50.7 | 0.252* |
| Savannah sparrow (O,M) | *Passerculus sandwichensis* | 2.9 | 2.5 | 4.1 | 0.610 |
| | | 9.6 | 6.2 | 19.1 | 0.325 |
| | | 14.6 | 9.8 | 37.6 | 0.261 |
| | | 19.1 | 16.4 | 80.3 | 0.204 |
| | | 17.0 | 13.0 | 67.7 | 0.192 |
| House finch (O,M) | *Carpodacus mexicanus* | 15.0 | 5.15 | 37 | 0.139 |
| Phainopepla (O, X and M) | *Phainopepla nitens* | 22.7 | 21.5 | 79.1 | 0.272 |
| Red-winged blackbird (C,H) | *Agelaius phoeniceus* | 30.2 | 36.8 | 99.5 | 0.370 |
| | | 30.2 | 12.1 | 82.9 | 0.146 |
| Wilson's storm petrel (C,SW) | *Oceanites oceanus* | 42.3 | 3.05 | 81 | 0.038 |
| | | 42.2 | 25.1 | 157 | 0.160* |
| Mockingbird (O,M) | *Mimus polyglottos* | 47.6 | 42.1 | 121 | 0.348 |
| Leach's storm petrel (C,SW) | *Oceanodroma leucorhoa* | 48 | 12.4 | 85.6 | 0.145* |
| Purple martin (C,M) | *Progne subis* | 47.7 | 33.0 | 183 | 0.180 |
| | | 50.3 | 30.7 | 143 | 0.215 |
| Starling (O,M) | *Sturnus vulgaris* | 85.0 | 62.7 | 231 | 0.271 |
| | | 78.7 | 77.7 | 246 | 0.316 |
| | | 74.1 | 79.7 | 327 | 0.244 |
| | | 76.9 | 80.3 | 272 | 0.295 |
| Gambel's quail (O,X) | *Callipepla gambelii* | 143 | 17.7 | 90.8 | 0.195* |
| Sand partridge (O,X) | *Ammoperdix heyi* | 156 | 17.9 | 122 | 0.147* |
| | | 206 | 24.3 | 150 | 0.162* |
| | | 209 | 17.8 | 172 | 0.103* |
| Kittiwake (C,SW) | *Rissa tridactyla* | 386 | 124 | 913 | 0.136* |
| Chukar partridge (O,X) | *Alectoris chukar* | 333 | 40.3 | 220 | 0.183* |
| | | 432 | 41.5 | 259 | 0.160* |
| | | 419 | 45.1 | 302 | 0.149* |

TABLE A3, continued

| GROUP Species | | Body mass, g | Water flux, ml/day | Energy metabolism, kJ/day | WEI, ml/kJ |
|---|---|---|---|---|---|
| BIRDS (cont.) | | | | | |
| Laysan albatross (C,SW) | *Diomedea immutabilis* | 3069 | 58.3 | 1450 | 0.040 |
| | | 3064 | 551 | 2160 | 0.255* |
| Jackass penguin (C,SW) | *Spheniscus demersus* | 3170 | 288 | 1950 | 0.148* |
| Adelie penguin (C,SW) | *Pygoscelis adeliae* | 3868 | 565 | 4000 | 0.141* |
| Giant petrel (C,SW) | *Macronectes giganteus* | 3583 | 469 | 4150 | 0.113* |
| | | 4505 | 586 | 4740 | 0.124* |
| Wandering albatross (C,SW) | *Diomedea exulans* | 7360 | 1330 | 2630 | 0.506* |
| | | 9440 | 1530 | 3970 | 0.395* |
| REPTILES | | | | | |
| Side-blotched lizard (C,X)[1] | *Uta stansburiana* | 3.3 | 0.15 | 0.570 | 0.263 |
| | | 2.8 | 0.07 | 0.432 | 0.162* |
| | | 3.0 | 0.05 | 0.222 | 0.225* |
| | | 2.5 | 0.03 | 0.123 | 0.244 |
| Sand lizard (C,X) | *Eremias lugubris* | 3.83 | 0.29 | 0.802 | 0.362 |
| Sand lizard (C,X) | *Eremias lineoocellata* | 3.27 | 0.16 | 0.545 | 0.294 |
| Orange-throated whiptail (C,X) | *Cnemidophorus hyperythrus* | 4.6 | 0.45 | 1.50 | 0.300 |
| | | 4.1 | 0.41 | 1.39 | 0.295* |
| | | 4.2 | 0.22 | 0.950 | 0.232* |
| | | 3.6 | 0.18 | 0.740 | 0.243* |
| Sand-diving lizard (O,X) | *Aporosaura anchietae* | 3.1 | 0.07 | 0.425 | 0.165* |
| | | 5.4 | 0.09 | 0.846 | 0.106 |
| Sagebrush lizard (C,X and M) | *Sceloporus graciosus* | 6.0 | 0.07 | 0.962 | 0.073 |
| | | 5.2 | 0.08 | 0.834 | 0.096 |
| | | 3.0 | 0.05 | 0.500 | 0.100 |
| Striped plateau lizard (C,M) | *Sceloporus virgatus* | 5.5 | 0.21 | 1.03 | 0.204* |
| | | 7.6 | 0.32 | 1.19 | 0.269* |
| | | 5.5 | 0.18 | 0.875 | 0.206* |
| | | 6.8 | 0.32 | 1.20 | 0.267* |
| | | 5.3 | 0.19 | 0.454 | 0.419* |
| Zebra-tailed lizard (C,X) | *Callisaurus draconoides* | 4.6 | 0.23 | 0.675 | 0.341 |
| | | 6.3 | 0.16 | 1.24 | 0.129 |
| | | 8.6 | 0.15 | 1.17 | 0.128 |

TABLE A3, continued

| GROUP Species | | Body mass, g | Water flux, ml/day | Energy metabolism, kJ/day | WEI, ml/kJ |
|---|---|---|---|---|---|
| REPTILES (cont.) | | | | | |
| Western fence lizard (C,M)[1] | *Sceloporus occidentalis* | 11.9 | 0.23 | 1.61 | 0.143 |
| | | 11.5 | 0.19 | 1.40 | 0.136 |
| | | 10.2 | 0.11 | 1.48 | 0.074 |
| | | 10.6 | 0.22 | 1.60 | 0.138 |
| | | 12.2 | 0.18 | 1.95 | 0.092 |
| Western whiptail (C,X) | *Cnemidophorus tigris* | 15.7 | 0.58 | 3.30 | 0.176* |
| Yarrow's spiny lizard (C,M) | *Sceloporus jarrovi* | 13.0 | 0.16 | 0.617 | 0.259* |
| | | 9.7 | 0.08 | 0.485 | 0.165 |
| | | 12.4 | 0.06 | 0.260 | 0.231* |
| | | 11.1 | 0.05 | 0.199 | 0.251* |
| | | 11.1 | 0.42 | 1.67 | 0.251 |
| | | 23.9 | 0.48 | 2.86 | 0.168 |
| | | 17.4 | 0.32 | 1.43 | 0.224 |
| | | 8.5 | 0.28 | 1.46 | 0.192 |
| Green lacerta (C,M) | *Lacerta viridis* | 25.5 | 3.06 | 5.84 | 0.524* |
| Lava lizard (C,H) | *Tropidurus albemarlensis* | 43.7 | 0.93 | 4.18 | 0.222* |
| | | 12.7 | 0.30 | 2.36 | 0.127* |
| Desert iguana (H,X) | *Dipsosaurus dorsalis* | 50.3 | 0.80 | 3.98 | 0.201 |
| | | 3.5 | 0.11 | 0.638 | 0.172* |
| Chuckwalla (H,X)[1] | *Sauromalus obesus* | 140 | 2.39 | 12.4 | 0.193 |
| | | 120 | 0.12 | 2.5 | 0.048 |
| Desert tortoise (H,X) | *Gopherus agassizii* | 950 | 11.4 | 39.5 | 0.289 |
| | | 950 | 7.6 | 39.5 | 0.192 |
| | | 950 | 11.4 | 34.6 | 0.329 |
| | | 950 | 0.95 | 11.7 | 0.081* |
| Marine iguana (H,H and SW)[1] | *Amblyrhynchus cristatus* | 69 | 2.15 | 4.85 | 0.443* |
| | | 1610 | 77.6 | 95.6 | 0.812* |
| Perentie (C,X) | *Varanus giganteus* | 5570 | 32.1 | 315 | 0.102 |
| | | 5570 | 167 | 659 | 0.253 |
| ARTHROPODS | | | | | |
| Desert beetle (X) | *Onymacris unguicularis* | 0.62 | 0.031 | 0.086 | 0.360 |
| Desert beetle (X)[1] | *Cryptoglossa verrucosa* | 0.60 | 0.0060 | 0.069 | 0.087* |
| | | 0.65 | 0.0020 | 0.021 | 0.095 |
| | | 0.69 | 0.0028 | 0.015 | 0.187* |
| | | 0.69 | 0.0200 | 0.110 | 0.182* |

TABLE A3, continued

| GROUP<br>Species | | Body mass, g | Water flux, ml/day | Energy metabolism, kJ/day | WEI, ml/kJ |
|---|---|---|---|---|---|
| ARTHROPODS (cont.) | | | | | |
| Darkling beetle (X)[1] | *Eleodes armata* | 0.93 | 0.033 | 0.115 | 0.287* |
| | | 0.77 | 0.0039 | 0.034 | 0.115 |
| | | 0.81 | 0.012 | 0.030 | 0.400* |
| | | 1.03 | 0.039 | 0.119 | 0.328 |
| | | 1.01 | 0.051 | 0.198 | 0.258* |
| Desert scorpion (X) | *Paruroctonus mesaensis* | 3.1 | 0.09 | 0.143 | 0.629* |
| | | 2.4 | 0.05 | 0.148 | 0.338* |
| Desert scorpion (X) | *Hadrurus arizonensis* | 3.5 | 0.094 | 0.605 | 0.155* |
| | | 3.5 | 0.137 | 0.575 | 0.238 |

[1] References for energy metabolism data listed in this table for each species are cited in Table A1, except for the following species: ground squirrel (Karasov 1981 and 1983b), jackrabbit (Shoemaker et al. 1976), howler monkey (Nagy and Milton 1979a), side-blotched lizard (Nagy 1983a), western fence lizard (Bennett and Nagy 1977), chuckwalla (Nagy and Shoemaker 1975), marine iguana (Nagy and Shoemaker 1984), and desert and darkling beetles (Cooper 1981).

* Maintaining constant body mass during measurements. (Others were gaining or losing mass, or mass change rates were not reported by authors.)

# Literature Cited

Adams, L.W., and T.J. Peterle.
1975. Retention of tritium in fresh-water marsh organisms. Journal of Great Lakes Research 1:1-9.

Adams, L.W., G.C. White, and T.J. Peterle.
1976. Tritium kinetics in a fresh-water marsh, pp. 96-103 in C.E. Cushing, Jr., ed., Radioecology and Energy Resources. Dowden, Hutchison and Ross, Stroudsburg, Pa.

Adams, N.J., C.R. Brown, and K.A. Nagy.
1986. Energy expenditure of free-ranging wandering albatrosses, *Diomedea exulans*. Physiological Zoology 59:583-591.

Adolph, E.F.
1949. Quantitative relations in the physiological constitutions of mammals. Science 109:579-585.

Afifi, A.A. and S.P. Azen.
1979. Statistical Analysis: A Computer-Oriented Approach, 2nd ed. Academic Press, New York.

Aggrey, E.K.
1982. Seasonal changes in water content and turnover in cattle, sheep, and goats grazing under humid tropical conditions in Ghana, pp. 133-142 in Use of Tritiated Water in Studies of Production and Adaptations in Ruminants. International Atomic Energy Agency, Vienna.

Alkon, P.U., A.A. Degen, B. Pinshow, and P.J. Shaw.
1985. Phenology, diet, and water turnover rates of Negev Desert chukars. Journal of Arid Environments 9:51-61.

Alkon, P.U., B. Pinshow, and A.A. Degen.
1982. Seasonal water turnover rates and body water volumes in desert chukars. Condor 84:332-337.

Alvarado, R.H.
1979. Amphibians, pp. 261-303 in G.M.O. Maloiy, ed., Comparative Physiology of Osmoregulation in Animals, Vol. 1. Academic Press, New York.

Anand, R.S., A.H. Parker, and H.R. Parker.
1966. Total body water and water turnover in sheep. American Journal of Veterinary Research 27:899-902.

Andersen, S.H., and E. Nielsen.
1983. Exchange of water between the harbor porpoise, *Phocoena phocoena*, and the environment. Experientia, Basel 39:52-53.

Anderson, R.A., and W.H. Karasov.
1981. Contrasts in energy intake and expenditure in sit-and-wait and widely foraging lizards. Oecologia 49:67-72.

Arai, K., Y. Kasida, and H. Takeda.
1975. Distribution of tritium in mice breeding with tritiated water. Journal of Radiation Research 16:82.

Argenzio, R.A., G.M. Ward, and J.E. Johnson.
1968. A study of body water spaces and water turnover in a ruminant. Journal of Animal Sciences 27:1121.

Arlian, L.G.
1975. Dehydration and survival of the European house-dust mite, *Dermatophagoides pteronyssinus*. Journal of Medical Entomology 12:437-442.

1979. Significance of passive sorption of atmospheric water vapor and feeding in water balance of the rice weevil, *Sitophilus oryzae*. Comparative Biochemistry and Physiology 62A:725-733.

Arlian, L.G., and I.A. Eckstrand.
1975. Water balance in the fruit fly, *Drosophila pseudoobscura*, and its ecological implications. Annals of the Entomological Society of America 68:827-832.

*Literature Cited*

Arlian, L.G., and T.E. Staiger.
1979. Water balance in the semi-aquatic beetle, *Peltodytes muticus*. Comparative Biochemistry and Physiology 62A:1041-1047.

Arlian, L.G., and M.M. Veselica.
1979. Water balance in insects and mites. Comparative Biochemistry and Physiology 64A:191-200.

Arlian, L.G., and G.W. Wharton.
1974. Kinetics of active and passive components of water exchange between the air and a mite, *Dermatophagoides farinae*. Journal of Insect Physiology 20:1063-1077.

Aschbacher, P.W., T.H. Kamal, and R.G. Gragle.
1965. Total body water estimations in dairy cattle, using tritiated water. Journal of Animal Science 24:430-433.

Azar, E., and S.J. Shaw.
1975. Effective body water half-life and total body water in rhesus and cynomolgus monkeys. Canadian Journal of Physiology and Pharmacology 53:935-939.

Bailey, C.B., and J.E. Lawson.
1981. Estimated water and forage intakes in nursing range calves. Canadian Journal of Animal Science 61:415-422.

Baker, N.F., A.L. Black, R.S. Anand, and R.A. Fisk.
1965. Body water turnover in cattle with parasitic gastroenteritus. Experimental Parasitology 17:271-276.

Bakker, H.R., S.D. Bradshaw, and A.R. Main.
1982. Water and electrolyte metabolism of the tammar wallaby, *Macropus eugenii*. Physiological Zoology 55:209-219.

Bartholomew, G.A.
1972. The water economy of seed-eating birds that survive without drinking, pp. 237-254 in K.H. Voous, ed., Proceedings of the XVth International Ornithological Congress. J.J. Brill, Leiden.

1982. Energy metabolism, pp. 46-93 in M.S. Gordon, G.A. Bartholomew, A.D. Grinnel, C.B. Jorgensen, and F.N. White, eds., Animal Physiology: Principles and Adaptations, 4th ed. Macmillan, New York.

Baverstock, P.R.
1975. Effect of variations in rate of growth on physiological parameters in the lizard, *Amphibolurus ornatus*. Comparative Biochemistry and Physiology 51A:619-631.

Bell, G.P., G.A. Bartholomew, and K.A. Nagy.
1986. The roles of energetics, water economy, foraging behavior, and geothermal refugia in the distribution of the bat, *Macrotus californicus*. Journal of Comparative Physiology 156:441-450.

Benjamin, R.W., M. Chen, A.A. Degen, N. Abdul Aziz, and M.J. Al Hadad.
1977. Estimation of the dry organic matter intake of young sheep grazing a dry Mediterranean pasture, and their maintenance requirements. Journal of Agricultural Science, Cambridge 88:513-520.

Benjamin, R.W., A.A. Degen, A. Brieghet, M. Chen, and N.H. Tadmor.
1975. Estimation of food intake of sheep grazing green pasture when no free water is available. Journal of Agricultural Science, Cambridge 85:403-407.

Bennett, A.F., and K.A. Nagy.
1977. Energy expenditure in free-ranging lizards. Ecology 58:697-700.

Bickler, P.E., and K.A. Nagy.
1980. Effects of parietalectomy on energy expenditure in free-ranging lizards. Copeia 1980:923-925.

Bird, P.R., J.J. Watson, J.W.D. Cayley, and J.F. Chin.
1980. Use of tritium-labeled water to estimate the intake of pasture by grazing cattle. Proceedings of the Australian Society of Animal Production 13:461.

Black, A.L., N.F. Baker, J.C. Bartley, T.E. Chapman, and R.W. Phillips.
1964. Water turnover in cattle. Science 144:876-878.

Bogdanov, K.M., M.I. Shalnov, and J.M. Stuckenberg.
1958. The use of tritium in the study of periodic biological phenomena: Isotopes in biochemistry and physiology, pp. 215-222 in part 2, Proceedings of the Second International Conference on the Peaceful Uses of Atomic Energy, Geneva.

Bohm, B.C., and N.F. Hadley.
1977. Tritium-determined water flux in the free-roaming desert tenebrionid beetle, *Eleodes armata*. Ecology 58:407-414.

Booth, D.T.
1987. Water flux in Malleefowl, *Leipoa ocellata* Gould (Megapodiidae). Australian Journal of Zoology 35:147-159.

Boxer, G.E., and D. Stetten, Jr.
1944. Studies in carbohydrate metabolism. II. The glycogenic response to glucose and lactate in the previously fasted rat. Journal of Biological Chemistry 155:237-242.

Bradford, D.F.
1974. Water stress of free-living *Peromyscus truei*. Ecology 55:1407-1414.

Bradshaw, S.D.
1978. Volume regulation in desert reptiles and its control by pituitary and adrenal hormones, pp. 38-59 in C.B. Jorgensen and E. Skadhauge, eds., Osmotic and Volume Regulation. Alfred Benzon Symposium XI, Munksgaard.

Bradshaw, S.D., J. Bradshaw, and F. Lachiver.
1976a. Quelques observations sur l'écophysiologie d'*Agama mutubilis* dans le sud Tunisien. Comptes Rendus des Seances de l'Academie des Sciences, Paris D282:93-96.

Bradshaw, S.D., T. Cheniti, and F. Lachiver.
1976b. Taux de renouvellement d'eau et balance hydrique chez deux rongeurs desertiques, *Meriones shawii* et *Meriones libycus*, étudiés dan leur environement naturel en Tunisie. Comptes Rendus des Seances de l'Academie des Sciences, Paris D282:481-484.

Bradshaw, S.D., H. Saint Girons, G. Naulleau, and K.A. Nagy.
1987. Material and energy balance of some captive and free-ranging reptiles in western France. Amphibia-Reptilia 8:129-142.

Brown, G.D., and J.J. Lynch.
1972. Some aspects of the water balance of sheep at pasture when deprived of drinking water. Australian Journal of Agricultural Research 23:669-684.

Brues, A.M., A.N. Stroud, and L. Rietz.
1952. Toxicity of tritium oxide to mice. Proceedings of the Society of Experimental Biology 79:174-176.

Buscarlet, L.A., and J. Proux.
1975. Étude a l'aide de l'eau tritiée du renouvellement de l'eau corporelle chez *Locusta migratoria migratorioides*. Comptes Rendus des Seances de l'Academie des Sciences, Paris D281:1409-1412.

Buscarlet, L.A., J. Proux, and R. Girster.
1978. Utilisation du double marquage $HT^{18}O$ dans une étude de bilan metabolique chez *Locusta migratoria migratorioides*. Journal of Insect Physiology 24:225-232.

Buss, D.H., and W.R. Voss.
1971. Evaluation of four methods for estimating the milk yield of baboons. Journal of Nutrition 101:901-910.

Butler, H.L., and J.H. Leroy.
1965. Observation of biological half-life of tritium. Health Physics 11:283-285.

Butte, N.F., C. Garza, E.O. Smith, and B.L. Nichols.
1983. Evaluation of the deuterium-dilution technique against the test-weighing procedure for the determination of breast-milk intake. American Journal of Clinical Nutrition 37:996-1003.

Calder, W.A. III.
1981. Scaling of physiological processes in homothermic animals. Annual Review of Physiology 43:301-322.

1984. Size, Function, and Life History. Harvard University Press, Cambridge, Mass.

Cameron, R.D., and J.R. Luick.
1972. Seasonal changes in total body water, extracellular fluid, and blood volume in grazing reindeer. Canadian Journal of Zoology 50:107-116.

Cameron, R.D., R.G. White, and J.R. Luick.
1976. Accuracy of the tritium water-dilution method for determining water flux in reindeer, *Rangifer tarandus*. Canadian Journal of Zoology 54:857-862.

Capen, R.L.
1972. Studies of water uptake in the euryhaline crab, *Rhithropanopeus harrisi*. Journal of Experimental Zoology 182:307-319.

Carrier, J.C., and D.H. Evans.
1972. Ion, water, and urea turnover rates in the nurse shark, *Ginglymostoma cirratum*. Comparative Biochemistry and Physiology 41A:761-764.

Carrier, J.C. and D.H. Evans
1973. Ion and water turnover in a fresh-water elasmobranch, *Potamotrygon* sp. Comparative Biochemistsry and Physiology 45A:667-670.

Chaouacha-Chekir, R.B., F. Lachiver, and T. Cheniti.
1983. Données préliminaires sur le taux de renouvellement d'eau chez un Gerbillidé desertique, *Psammomys obesus*, étudié dans son environment naturel en Tunisie. Mammalia 47:543-548.

Chapman, T.E., and A.L. Black.
1967. Water turnover in chickens. Poultry Science 46:761.

Chapman, T.E., and L.Z. McFarland.
1971. Water turnover in coturnix quail, with individual observations on a burrowing owl, Petz conure, and vulturine fish eagle. Comparative Biochemistry and Physiology 39A:653-656.

Chapman, T.E., and D. Mihai.
1972. Influences of sex and egg production on water turnover in chickens. Poultry Science 51:1252-1256.

Chemama, R., S. Apelgot, G. Rudali, M. Frilley, and E. Coezy.
1972. Courbes d'épuration de la radioactivité de divers tissus provenant de souris réfractaires ou prédisposées au cancer de la mamelle. II. Après administration d'eau tritiée. Bulletin du Cancer, Paris 59:207-224.

Che Mat, C.R.B., and W.T.W. Potts.
1985. Water balance in *Crangon vulgaris*. Comparative Biochemistry and Physiology 82A:705-710.

Chiswell, W.D., and G.H. Dancer.
1969. Measurement of tritium concentration in exhaled water vapor as a means of estimating body burdens. Health Physics 17:331-334.

Church, R.L.
1966. Water exchanges of the California vole, *Microtus californicus*. Physiological Zoology 39:326-340.

Cohen, A.C., R.B. March, and J.D. Pinto.
1981. Water relations of the desert blister beetle, *Cysteodemus armatus* (LeConte) (Coleoptera: Meloidae). Physiological Zoology 54:179-187.

Congdon, J.D.
1977. Energetics of the montane lizard (*Sceloporus jarrovi*): A measure of reproductive effort. Ph.D. dissertation, Arizona State University, Tempe.

Congdon, J.D., R.E. Ballinger, and K.A. Nagy.
1979. Energetics, temperature, and water relations in winter-aggregated *Sceloporus jarrovi* (Sauria: Iguanidae). Ecology 60:30-35.

Congdon, J.D., and D.W. Tinkle.
1982. Energy expenditure in free-ranging sagebrush lizards (*Sceloporus graciosus*). Canadian Journal of Zoology 60:1412-1416.

Congdon, J.D., L.J. Vitt, R.C. Van Loben Sels, and R.D. Ohmart.
1982. The ecological significance of water flux rates in arboreal desert lizards of the genus *Urosaurus*. Physiological Zoology 55:317-322.

Cooper, P.D.
1982. Water balance and osmoregulation in a free-ranging tenebrionid beetle, *Onymacris unguicularis*, of the Namib Desert. Journal of Insect Physiology 28:737-742.

1983. Validation of the doubly labeled water ($H^3H^{18}O$) method for measuring water flux and energy metabolism in tenebrionid beetles. Physiological Zoology 56:41-46.

1985. Seasonal changes in water budgets in two free-ranging tenebrionid beetles, *Eleodes armata* and *Cryptoglossa verrucosa*. Physiological Zoology 58:458-472.

Costa, D.P.
1982. Energy, nitrogen, and electrolyte flux and sea water drinking in the sea otter, *Enhydra lutris*. Physiological Zoology 55:35-44.

1984. Assessment of the impact of the California sea lion and northern elephant seal on commercial fisheries. California Sea Grant College Program Biennial Report, 1982-84. Institute of Marine Resources, University of California, San Diego.

Costa, D.P., and R.L. Gentry.
1986. Free-ranging energetics of northern fur seals, pp. 79-101 in R.L. Gentry and G.L. Kooyman, eds., Fur Seals: Maternal Strategies on Land and at Sea, Princeton, N.J.

Costa, D.P., B.J. LeBoeuf, A.C. Huntley, and C.L. Ortiz.
In press. The energetics of lactation in the northern elephant seal. Journal of Zoology, London.

Costa, D.P., and P.A. Prince.
1987. Foraging energetics of grey-headed albatrosses *Diomedea chrysostoma* at Bird Island, South Georgia. Journal of Zoology, 129:149-158.

Coward, W.A., T.J. Cole, H. Gerber, S.B. Roberts, and I. Fleet.
1982. Water turnover and the measurement of milk intake. Pfluegers Archiv: European Journal of Physiology 393:344-347.

Coward, W.A., M.B. Sawyer, R.G. Whitehead, A.M. Prentice, and J. Evans.
1979. New method for measuring milk intakes in breast-fed babies. Lancet 1979:13-14.

Daily, C., and H.B. Haines.
1981. Evaporative water loss and water turnover in chronically and acutely water-restricted spiny mice (*Acomys cahirinus*). Comparative Biochemistry and Physiology 68A:349-354.

Davis, R.W., G.L. Kooyman, and J.P. Croxall.
1983. Water flux and estimated metabolism of free-ranging gentoo and macaroni penguins at South Georgia. Polar Biology 2:41-46.

Davis, T.P., M.K, Yousef, and H.D. Johnson.
1978. Partition of body fluids in the burro. Journal of Wildlife Management 42:923-925.

Dawson, T.J., M.J.S. Denny, E.M. Russell, and B. Ellis.
1975. Water usage and diet preferences of free-ranging kangaroos, sheep, and feral goats in the Australian arid zone during summer. Journal of Zoology 177:1-23.

Dawson, T.J., R.M. Herd, and E. Skadhauge.
1983. Water turnover and body water distribution during dehydration in a large arid-zone bird, the emu, *Dromaius novae-hollandiae*. Journal of Comparative Physiology 153:235-240.

Dawson, T.J., D. Read, E.M. Russell, and R.M. Herd.
1984. Seasonal variation in daily activity patterns, water relations and diet of emus, *Dromaius novae-hollandiae*. The Emu 84:93-102.

Deavers, D.R., and J.W. Hudson.
1977. Effect of cold exposure on water requirements of three species of small mammals. Journal of Applied Physiology 43:121-125.

Deavers, D.R., and X.J. Musacchia.
1976. Water turnover, total body water, and metabolism of hibernating and cold-acclimated normothermic thirteen-lined ground squirrels. Federation Proceedings 35:639.

Degabriele, R., C.J.F. Harrop, and T.J. Dawson.
1978. Water metabolism of the koala, (*Phascolarctos cinereus*), pp. 163-172 in G.G. Montgomery, ed., The Ecology of Arboreal Folivores. Proceedings of the National Zoological Park Symposia, no. 1. Smithsonian Institution Press, Washington, D.C.

Degani, G.
1982. Water balance of salamanders (*Salamandra salamandra* (L.)) from different habitats. Amphibia-Reptilia 2:309-314.

1985. Water balance and body fluids of *Salamandra salamandra* (L.) in their natural habitats in summer and winter. Comparative Biochemistry and Physiology 82A:479-482.

Degen, A.A.
1977. Fat-tailed awassi and German mutton merino sheep under semi-arid conditions. I. Total body water, its distribution, and water turnover. II. Total body water and water turnover during pregnancy and lactation. Journal of Agricultural Science, Cambridge 88:693-698, 88:699-704.

Degen, A.A., M. Kam, A. Hazan, and K.A. Nagy.
1986. Energy expenditure and water flux in three sympatric desert rodents. Journal of Animal Ecology 55:421-429.

Degen, A.A., B. Pinshow, and P.U. Alkon.
1982. Water flux in chukar partridges (*Alectoris chukar*) and a comparison with other birds. Physiological Zoology 55:64 71.

1983a. Summer water turnover rates in free-living chukars and sand partridges in the Negev Desert. The Condor 85:333-337.

Degen, A.A., B. Pinshow, P.U. Alkon, and H. Arnon.
1981. Tritium labelled water for estimating total body water and water turnover rate in birds. Journal of Applied Physiology 51:R1183-R1188.

Degen, A.A., B. Pinshow, and P.J. Shaw.
1983b. Must desert chukars (*Alectoris chukar sinaica*) drink water? Water influx and body mass changes in response to dietary water content. The Auk 101:47-52.

Degen, A.A., and B.A. Young.
1981. Effect of air temperature and feed intake on live weight and water balance in sheep. Journal of Agricultural Science, Cambridge 96:493-496.

Denny, M.J.S., and T.J. Dawson.
1973. A field technique for studying water metabolism of large mammals. Journal of Wildlife Management 37:574-578.

1975. Comparative metabolism of tritiated water by macropodid marsupials. American Journal of Physiology 228:1794-1799.

Depocas, F., J.S. Hart, and H.D. Fisher.
1971. Sea-water drinking and water flux in starved and in fed harbor seals, *Phoca vitulina*. Canadian Journal of Physiology and Pharmacology 49:53-62.

Devine, T.L.
1977. Incorporation of tritium from water into tissue components of the book louse, *Liposulis bostrychophilus*. Journal of Insect Physiology 23:1315-1322.

Devine, T.L., and G.W. Wharton.
1973. Kinetics of water exchange between a mite, *Laelaps echidnina*, and the surrounding air. Journal of Insect Physiology 19:243-254.

Dixon, W.J. and F.J. Massey.
1969. Introduction to Statistical Analysis, 3rd ed. McGraw-Hill, New York.

Doreau, M., G. Dussap, and H. Dubroeucq.
1980. Estimation of nursing mare milk production using deuterium oxide as a tracer of offspring body water. Reproduction, Nutrition, Developpement 20:1883-1892.

Dove, H., and M. Freer.
1979. The accuracy of tritiated water turnover rate as an estimate of milk intake in lambs. Australian Journal of Agricultural Research 30:725-739.

Duff, D.W., and W.R. Fleming.
1972a. Sodium metabolism of the fresh-water cyprinodont, *Fundulus catenatus*. Journal of Comparative Physiology 80:179-189.

Duff, D.W. and W.R. Fleming
1972b. Some aspects of sodium balance in the fresh-water cyprinodont, *Fundulus olivaceus*. Journal of Comparative Physiology 80:191-199.

Dunn, O.J., and V.A. Clark.
1974. Applied Statistics: Analysis of Variance and Regression. John Wiley, New York.

Dunson, W.A.
1978. Role of the skin in sodium and water exchange of aquatic snakes placed in sea water. American Journal of Physiology 235:R151-R159.

1980. The relation of sodium and water balance to survival in sea water of estuarine and fresh water races of the snakes *Nerodia fasciata*, *Nerodia sipedon*, and *Nerodia valida*. Copeia 1980:268-280.

1985. Effect of water salinity and food salt content on growth and sodium efflux of hatching diamondback terrapins (*Malaclemys*). Physiological Zoology 58:736-747.

Dunson, W.A., R.K. Packer, and M.K. Dunson.
1971. Ion and water imbalance in normal and mutant-imbalanced (ff) embryos in the axolotl (*Ambystoma mexicanum*). Comparative Biochemistry and Physiology 40A:319-337.

Dunson, W.A., and G.D. Robinson.
1976. Sea-snake skin, permeable to water but not to sodium. Journal of Comparative Physiology 108:303-311.

Dupre, R.K.
1983. A comparison of the water relations of the hispid cotton rat, *Sigmodon hispidus*, and the prairie vole, *Microtus ochrogaster*. Comparative Biochemistry and Physiology 75A:659-663.

Edney, E.B.
1977. Water Balance in Land Arthropods. Springer-Verlag, New York.

Edney, E.B., and K.A. Nagy.
1976. Water balance and excretion. pp. 106-132 in J. Bligh, J.L. Cloudsley-Thompson, and A.G. Macdonald, eds., Environmental Physiology of Animals. Blackwell Scientific Publications, Oxford, England.

*Literature Cited*

Ehleringer, J.R., E.-D. Schulze, H. Ziegler, O.L. Lange, G.D. Farquhar, and I.R. Cowar.
1985. Xylem-tapping mistletoes: Water or nutrient parasites? Science 227:1479-1481.

Ehrenfeld, J., and J. Isaia.
1974. The effect of ligaturing the eyestalks on water and ion permeabilities of *Astacus leptodactylus*. Journal of Comparative Physiology 93:105-115.

Elford, B.C.
1972. Permeation and distribution of deuterated and tritiated water in smooth and striated muscle. Journal of Physiology, London 211:73-92.

El Hadi, H.M., and Y.M. Hassan.
1982. Seasonal changes in water metabolism of Sudan Desert sheep and goats, pp. 103-115 in Use of Tritiated Water in Studies of Production and Adaptations in Ruminants. International Atomic Energy Agency, Vienna.

Ellingsen, I.J.
1975. Permeability to water in different adaptive phases of the same instar in the American house-dust mite. Acarologia 17:734-744.

Etzion, Z., N. Meyerstein, and R. Yagil.
1984. Tritiated water metabolism during dehydration and rehydration in the camel, *Camelus dromedarius*. Journal of Applied Physiology 56:217-220.

Evans, D.H.
1967. Sodium, chloride, and water balance of the intertidal teleost, *Xiphister atropurpureus*. III. The role of simple diffusion, exchange diffusion, osmosis, and active transport. Journal of Experimental Biology 47:525-534.

1969a. Sodium chloride and water balance of the intertidal teleost *Pholis gunnellus*. Journal of Experimental Biology 50:179-190.

1969b. Studies on the permeability to water of selected marine, fresh-water, and euryhaline teleosts. Journal of Experimental Biology 50:689-703.

1979. Fish, pp. 305-390 in G.M.O. Maloiy, ed., Comparative Physiology of Osmoregulation in Animals, Vol. 1. Academic Press, New York.

Faichney, G.J., and R.C. Boston.
1985. Movement of water within the body of sheep fed at maintenance under thermoneutral conditions. Australian Journal of Biological Sciences 38:85-94.

Fallot, P., A. Aaberhardt, and J. Masson.
1957. Methode de dosage de l'eau tritiée et ses applications en clinique humaine. International Journal of Applications of Radioactive Isotopes 1:237-245.

Fiala, K.L., and J.D. Congdon.
1983. Energetic consequences of sexual size dimorphism in nestling red-winged blackbirds. Ecology 64:642-647.

Finch, V.A., and J.M. King.
1982. Energy-conserving mechanisms as adaptation to undernutrition and water deprivation in the African zebu, pp. 167-178 in Use of Tritiated Water in Studies of Production and Adaptations in Ruminants. International Atomic Energy Agency, Vienna.

Fletcher, C.R.
1974. Volume regulation in *Nereis diversicolor*. I. The steady state. II. The effect of calcium. Comparative Biochemistry and Physiology 47A:1199-1214, 47A:1215-1220.

1978. Osmotic and ionic regulation in the cod, *Gadus callarias*. I. Water balance. Journal of Comparative Physiology 124B:149-156.

Foster, M.A.
1969. Ionic and osmotic regulation in three species of *Cottus* (Cottidae, Teleostei). Comparative Biochemistry and Physiology 30:751-759.

Foy, J.M.
1964. The biological half-life of tritiated water in the mouse, rat, guinea pig, and rabbit, and the effect of climate and saline drinking on the biological half-life of tritiated water in the rat. Journal of Cellular and Comparative Physiology 64:279-282.

Foy, J.M., and H. Schnieden.
1960. Estimation of total body water (virtual tritium space) in the rat, cat, rabbit, guinea pig, and man, and of the biological half-life of tritium in man. Journal of Physiology 154:169-176.

Gabrielsen, G.W., F. Mehlum, and K.A. Nagy.
1987. Daily energy expenditure and energy utilization of free-ranging black-legged kittiwakes. The Condor 89:126-132.

Gaebler, O.H., and H.C. Choitz.
1964.   Studies of body water and water turnover determined with deuterium oxide added to food. Clinical Chemistry 10:13-20.

Gettinger, R.D.
1983.   Use of doubly labeled water ($^3HH^{18}O$) for determination of $H_2O$ flux and $CO_2$ production by a mammal in a humid environment. Oecologia 59:54-57.

1984.   Energy and water metabolism of free-ranging pocket gophers, *Thomomys bottae*. Ecology 65:740-751.

Gettinger, R.D., W.W. Weathers, and K.A. Nagy.
1985.   Energetics of free-living nestling house finches: Measurements with doubly labeled water. The Auk 102:643-644.

Gilles, R.
1979.   Mechanisms of Osmoregulation in Animals: Maintenance of Cell Volume. John Wiley, New York.

Gleason, J.D., and I. Friedman.
1970.   Deuterium: Natural variations used as a biological tracer. Science 169:1085-1086.

Goldstein, D.L., and K.A. Nagy.
1985.   Resource utilization by desert quail: Time and energy, food and water. Ecology 66:378-387.

Golightly, R.T., Jr., and R.D. Ohmart.
1984.   Water economy of two desert canids, coyote and kit fox. Journal of Mammalogy 65:51-58.

Green, B.
1972.   Water losses of the sand goanna (*Varanus gouldii*) in its natural environment. Ecology 53:452-457.

Green, B., J. Anderson, and T. Whateley.
1984.   Water and sodium turnover and estimated food consumption in free-living lions (*Panthera leo*) and spotted hyaenas (*Crocuta crocuta*). Journal of Mammalogy 65:593-599.

Green, B., and J.D. Dunsmore.
1978.   Turnover of tritiated water and Na-22 in captive rabbits (*Oryctolagus cuniculus*). Journal of Mammalogy 59:12-17.

Green, B., J.D. Dunsmore, H. Bults, and K. Newgraim.
1978. Turnover of sodium and water by free-living rabbits, *Oryctolagus cuniculus*. Australian Wildlife Research 5:93-99.

Green, B., and I. Eberhard.
1979. Energy requirements and sodium and water turnovers in two captive marsupial carnivores: the Tasmanian devil, *Sarcophilus harrisii*, and the native cat, *Dasyurus viverrinus*. Australian Journal of Zoology 27:1-8.

1983. Water and sodium intake and estimated food consumption in free-living eastern quolls, *Dasyurus viverrinus*. Australian Journal of Zoology 31:871-880.

Green, B., D. King, and H. Butler.
1986. Water, sodium, and energy turnover in free-living perenties, *Varanus giganteus*. Australian Wildlife Research 13:589-596.

Grenot, C., M. Pascal, L.A. Buscarlet, J.M. Francaz, and M. Sellami.
1984. Water and energy balance in the water vole (*Arvicola terrestris* Sherman) in the laboratory and in the field (Haut-Doubs, France). Comparative Biochemistry and Physiology 78A:185-196.

Grenot, C., and V. Serrano.
1979. Vitesse de renouvellement d'eau chez cinq especes de Rongeurs deserticoles et sympatriques etudiées a la saison seche dans leur milieu naturel (Desert de Chihuahua, Mexique). Comptes Rendus des Seances de l'Academie des Sciences, Paris D288:104.

Grigg, G.C., L.E. Taplin, B. Green, and P. Harlow.
1986. Sodium and water fluxes in free-living *Crocodylus porosus* in marine and brackish conditions. Physiological Zoology 59:240-253.

Grubbs, D.E.
1980. Tritiated water turnover in free-living desert rodents. Comparative Biochemistry and Physiology 66A:89-98.

Haglund, B., and S. Lovtrup.
1966. The influence of temperature on the water exchange in amphibian eggs and embryos. Journal of Cell Physiology 67:355-360.

Haines, H., C. Ciskowski, and V. Harms.
1973a. Acclimation to chronic water restriction in the wild house mouse, *Mus musculus*. Physiological Zoology 46:110-128.

Haines, H., B. Howard, and C. Setchell.
1973b. Water content and distribution of tritiated water in tissues of Australian desert rodents. Comparative Biochemistry and Physiology 45A:787-792.

Haines, H., W.V. Macfarlane, C. Setchell, and B. Howard.
1974. Water turnover and pulmocutaneous evaporation of Australian desert dasyurids and murids. American Journal of Physiology 227:958-963.

Haley, H.B., I.S. Edelman, D.S. Frederickson, and F.D. Moore.
1953. Water balance and electrolytes: The effect of thyroid disease on water turnover rate in the human subject. Surgical Forum 4:543-547.

Haley, H.B., I.S. Edelman, and F.D. Moore.
1951. The turnover rate of body water in the human. Surgical Forum 2:595.

Hannan, J.V., and D.H. Evans.
1973. Water permeability in some euryhaline decapods and *Limulus polyphemus*. Comparative Biochemistry and Physiology 44A:1199-1213.

Hatch, F.T., and J.A. Mazrimas.
1972. Tritiation of animals from tritiated water. Radiation Research 50:339-357.

Haywood, G.P.
1974. The exchangeable ionic space and salinity effects upon ion, water, and urea turnover rates in dogfish, *Poroderma africanum*. Marine Biology 26:69-75.

Helversen, O.V., and H.U. Reyer.
1984. Nectar intake and energy expenditure in a flower-visiting bat. Oecologia 63:178-184.

Hevesy, G., and R. Hofer.
1934. Elimination of water from the human body. Nature, London 134:879.

Hewitt, S., J.F. Wheldrake, and R.V. Baudinette.
1981. Water balance and renal function in the Australian desert rodent, *Notomys alexis*: The effect of diet on water turnover rate, glomerular filtration rate, renal plasma flow, and renal blood flow. Comparative Biochemistry and Physiology 68A:405-410.

Hinds, D.S., and R.E. MacMillen.
1985. Scaling of energy metabolism and evaporative water loss in heteromyid rodents. Physiological Zoology 58:282-298.

Holleman, D.F., and R.A. Dieterich.
1973. Body water content and turnover in several species of rodents, as evaluated by the tritiated water method. Journal of Mammalogy 54:456-465.

Holleman, D.F., R.G. White, and D.D. Feist.
1982. Seasonal energy and water metabolism in free-living Alaskan voles. Journal of Mammalogy 63:293-296.

Holleman, D.F., R.G. White, and J.R. Luick.
1975. New isotope methods of estimating milk intake and yield. Journal of Dairy Science 58:1814-1821.

Hudson, J.W., and J.A. Rummel.
1966. Water metabolism and temperature regulation of the primitive heteromyids, *Liomys salvani* and *Liomys irroratus*. Ecology 47:345-354.

Hughes, M.R., J.R. Roberts, and B.R. Thomas.
1987. Total body water and its turnover in free-living nestling glaucous-winged gulls with a comparison of body water and water flux in avian species with and without salt glands. Physiological Zoology 60:481-491.

Hui, C.A.
1981. Sea water consumption and water flux in the common dolphin, *Delphinus delphis*. Physiological Zoology 54:430-440.

Hulbert, A.J., and T.J. Dawson.
1974. Water metabolism in perameloid marsupials from different environments. Comparative Biochemistry and Physiology 47A:617-633.

Hulbert, A.J., and G. Gordon.
1972. Water metabolism of the bandicoot *Isoodon macrourus* Gould in the wild. Comparative Biochemistry and Physiology 41A:27-34.

Hulbert, A.J., and T.R. Grant.
1983. A seasonal study of body condition and water turnover in a free-living population of platypuses, *Ornithorhynchus anatinus* (Monotremata). Australian Journal of Zoology 31:109-116.

Hume, I.D., and A. Dunning.
1979. Nitrogen and electrolyte balance in the wallabies *Thylogale thetis* and *Macropus eugenii* when given saline drinking water. Comparative Biochemistry and Physiology 63A:135-139.

Hutchinson, D.L., A.A. Plentl, and H.C. Taylor, Jr.
1954. The total body water and the water turnover in pregnancy studies with deuterium oxide as isotopic tracer. Journal of Clinical Investigation 33:235.

Isaia, J.
1972. Comparative effects of temperature on the sodium and water permeabilities of the gills of a stenohaline freshwater fish (*Carassius auratus*) and a stenohaline marine fish (*Serranus scriba, Serranus cabrilla*). Journal of Experimental Biology 57:359-366.

Jones, G.B., B.J. Potter, and C.S.W. Reid.
1970. The effect of saline water ingestion on water turnover rates and tritiated water space in sheep. Australian Journal of Agricultural Research 21:927-932.

Kalanidhi, A.P., K.K. Saxena, V.P. Shukla, and S.K. Ranjhan.
1980. Water turnover rate in native and cross-bred sheep. Indian Journal of Animal Sciences 50:179-181.

Kam, M., A.A. Degen, and K.A. Nagy.
1987. Seasonal energy, water and food consumption of Negev chukars and sand partridges. Ecology 68:1029-1037.

Kamal, T.H.
1982. Water turnover rate and total body water as affected by different physiological factors under Egyptian environmental conditions, pp. 143-154 in Use of Tritiated Water in Studies of Production and Adaptations in Ruminants. International Atomic Energy Agency, Vienna.

Kamal, T.H., and H.D. Johnson.
1971. Total body solids loss as a measure of a short-term heat stress in cattle. Journal of Animal Science 32:306.

Kamal, T.H., O. Shehata, and I.M. Elbanna.
1972. Effect of heat and water restriction on water metabolism and body fluid compartments in farm animals, pp. 95-100 in Isotope Studies on the Physiology of Domestic Animals. International Atomic Energy Agency, Vienna.

Kamis, A.B., and N.B. Latif.
1981. Turnover and total body water in the macaque, *Macaca fascicularis*, and the gibbon, *Hylobates lar*. Comparative Biochemistry and Physiology 70A:45-46.

Karasov, W.H.
1981. Daily energy expenditure and the cost of activity in a free living mammal. Oecologia 51:253-259.

1983a. Water flux and water requirement in free-living antelope ground squirrels (*Ammospermophilus leucurus*). Physiological Zoology 56:94-105.

1983b. Wintertime energy conservation by huddling in antelope ground squirrels, *Ammospermophilus leucurus*. Journal of Mammalogy 64:341-345.

Karasov, W.H., and R.A. Anderson.
1984. Interhabitat differences in energy acquisition and expenditure in a lizard. Ecology 64:235-247.

Kennedy, P.M., and G.E. Heinsohn.
1974. Water metabolism of two marsupials: The brush-tailed possum, *Trichosurus vulpecula*, and the rock wallaby, *Petrogale inornata*, in the wild. Comparative Biochemistry and Physiology 47A:829-834.

Kennedy, P.M., and W.V. Macfarlane.
1971. Oxygen consumption and water turnover of the fat-tailed marsupials *Dasycercus cristicauda* and *Sminthopsis crassicaudata*. Comparative Biochemistry and Physiology 40A:723-732.

King, J.M.
1979. Game domestication for animal production in Kenya field studies of the body water turnover of game and livestock. Journal of Agricultural Science 93:71-80.

King, J.M., G.P. Kingaby, J.G. Colvin, and B.R. Heath.
1975. Seasonal variation in water turnover by oryx and eland on the Galana Game Ranch Research Project. East African Wildlife Journal 13:287-296.

King, J.M., P.O. Nyamora, M.R. Stanley-Price, and B.R. Heath.
1978. Game domestication for animal production in Kenya: Prediction of water intake from tritiated water turnover. Journal of Agricultural Science 91:513-522.

King, W.W., and N.F. Hadley.
1979. Water flux and metabolic rates of free-roaming scorpions, using the doubly labeled water technique. Physiological Zoology 52:176-189.

*Literature Cited* 151

Kitchener, D.J.
1970. Aspects of the response of the quokka to environmental stress. Ph.D. dissertation, University of Western Australia, Perth.

Knox, K.L., A. Chappell, J.A. Gibbs, D.N. Hyder, and R.E. Bement.
1970. Sampling methods for water kinetics. Journal of Dairy Science 53:1279-1282.

Knox, K.L., J.G. Nagy, and R.D. Brown.
1969. Water turnover in mule deers. Journal of Wildlife Management 33:389-393.

Knulle, W., and T.L. Devine.
1972. Evidence for active and passive components of sorption of atmospheric water vapor by larvae of the tick, *Dermacentor variabilis*. Journal of Insect Physiology 18:1653-1664.

Kooyman, G.L., R.W. Davis, J.P. Croxall, and D.P. Costa.
1982. Diving depths and energy requirements of king penguins. Science 217:726-727.

Krogh, A.
1939. Osmotic Regulation in Aquatic Animals. Cambridge University Press, Cambridge, England.

Lachiver, F., T. Cheniti, S.D. Bradshaw, J.L. Berthier, and F. Petter.
1978. Field studies in southern Tunisia on water turnover and thyroid activity in two species of *Meriones*, pp. 81-84 in F. Assenmacher and D.S. Founer, eds., Environmental Endocrinology. Springer-Verlag, New York.

Lahlou, B., and A. Giordan.
1970. Hormonal control of water balance and exchange in fresh-water teleostean *Carassius auratus*, intact and hypophysectomized. General and Comparative Endocrinology 14:491-509.

Lahlou, B., and W.H. Sawyer.
1969a. Influence de l'hypophysectomie sur le renouvellement de l'eau interne (etudiée a l'aide de l'eau tritiée) chez le poisson rouge, *Carassius auratus* L. Comptes Rendus des Seances de l'Academie des Sciences, Paris D268:725-728.

1969b. Adaptation des échanges d'eau en fonction de la salinité externe chez *Opsanus tau*, teleostéen agiomerulaire euryhalin. Journal de Physiologie, Paris 61:143.

Lambert, B.E., and J.R. Clifton.
1967. Radiation doses resulting from the administration of tritiated folic acid and tritiated water to the rat. British Journal of Radiology 40:56.

Lasiewski, R.C., and W.R. Dawson.
1967. A re-examination of the relation between standard metabolic rate and body weight in birds. The Condor 69:13-23.

Leader, J.P.
1972. Osmoregulation in the larva of the marine caddis fly, *Philanisus plebeius* (Walk.) (Trichoptera). Journal of Experimental Biology 57:821-838.

Lee, J.S., and N. Lifson.
1960. Measurement of total energy and material balance in rats by means of doubly labeled water. American Journal of Physiology 199:238-242.

Lemire, M., C. Grenot, and R. Vernet.
1979. La balance hydrique d'*Uromastix acanthinurus* Bell (Sauria, Agamidae) au Sahara dans des conditions seminaturelles. Comptes Rendus des Seances de l'Academie Sciences, Paris D288:359-362.

Lifson, N., G.B. Gordon, R. McClintock.
1955. Measurement of total carbon dioxide production by means of doubly labelled water. Journal of Applied Physiology 7:704-710.

Lifson, N., and J.S. Lee.
1961. Estimation of material balance of totally fasted rats by doubly labeled water. American Journal of Physiology 200:85-88.

Lifson, N., and R. McClintock.
1966. Theory of use of the turnover rates of body water for measuring energy and material balance. Journal of Theoretical Biology 12:46-74.

Little, W.S., and N. Lifson.
1975. Validation study of the $D_2O$-18 method for determination of $CO_2$ output of the eastern chipmunk (*Tamias striatus*). Comparative Biochemistry and Physiology 50A:55-56.

Lockwood, A.P.M.
1959. The osmotic and ionic regulation of *Asellus aquaticus* (L.). Journal of Experimental Biology 36:546-555.

1961. The urine of *Gammarus duebeni* and *G. pulex*. Journal of Experimental Biology 38:647-658.

Lockwood, A.P.M., and C.B.E. Inman.
1973. Changes in the apparent permeability to water at moult in the amphipod *Gammarus duebeni* and the isopod *Idotea linearis*. Comparative Biochemistry and Physiology 44A:943-952.

Lockwood, A.P.M., C.B.E. Inman, and T.H. Courtenay.
1973. The influence of environmental salinity on the water fluxes of the amphipod crustacean *Gammarus duebeni*. Journal of Experimental Biology 58:137-148.

Longhurst, W.M., N.F. Baker, G.E. Connolly, and R.A. Fisk.
1970. Total body water and water tunover in sheep and deer. American Journal of Veterinary Research 31:673-677.

Lopez, G.A., R.W. Phillips, and C.F. Nockels.
1973a. The effect of age on water metabolism in hens. Proceedings of the Society of Experimental Biology and Medicine 143:545-547.

1973b. Body water kinetics in vitamin A-deficient chickens. Proceedings of the Society of Experimental Biology and Medicine 144:54-55.

Loretz, C.
1979. Water exchange across fish gills: The significance of tritiated-water flux measurements. Journal of Experimental Biology 79:147-162.

Lotan, R.
1969. Sodium chloride and water balance in the euryhaline teleost, *Aphanius dispar* (Ruppell) (Cyprinodontidae). Zeitschrift für Vergleichende Physiologie 65:455-462.

Louw, G.N., and M.K. Seely.
1982. Ecology of Desert Organisms. Longman, New York.

Lynch, J.J., G.D. Brown, P.F. May, and J.B. Donnelly.
1972. The effect of withholding drinking water on wool growth and lamb production of grazing merino sheep in a temperate climate. Australian Journal of Agricultural Research 23:659-668.

Macfarlane, N.A.A., and J. Maetz.
1974. Effects of hypophysectomy on sodium and water exchanges in the euryhaline flounder, *Platychthys flesus*. General and Comparative Endocrinology 22:77-89.

Macfarlane, W.V.

1965. Water metabolism of desert ruminants, pp. 191-199 in A.K. McIntyre and D.R. Curtis, eds., Studies in Physiology. Springer-Verlag, New York.

1968. Comparative function of ruminants in hot environments, pp. 264-276 in E.S.E. Hafez, ed., Adaptations of Domestic Animals. Lea and Febiger, Philadelphia.

1969. The water economy of desert aboriginals in summer. Journal of Physiology, London 205:13P-14P.

1975. Ecophysiology of water and energy in desert marsupials and rodents, pp. 389-396 in I. Prakash and P.K. Ghosh, eds., Rodents in Desert Environments. Dr. Junk Publishers, The Hague.

1976. Ecophysiological hierarchies. Israel Journal of Medical Science 12:723-731.

Macfarlane, W.V., C.H.S. Dolling, and B. Howard.

1966a. Distribution and turnover of water in merino sheep selected for high wool production. Australian Journal of Agricultural Research 17:491-502.

Macfarlane, W.V., and B. Howard.

1966a. Water turnover: Mammals, pp. 505-506 in P.L. Altman and D.S. Dittmer, eds., Environmental Biology. Federation of the American Societies of Experimental Biology, Bethesda, Md.

1966b. Water content and turnover of identical twin *Bos indicus* and *B. taurus* in Kenya. Journal of Agricultural Science 66:297-302.

1970. Water in the physiological ecology of ruminants, pp. 362-374 in A.T. Phillipson, ed., The Physiology of Digestion and Nutrition in the Ruminant. Oriel, Newcastle, Conn.

1972. Comparative water and energy economy of wild and domestic mammals. Symposia of the Zoological Society of London 31:261-296.

Macfarlane, W.V., B. Howard, and B.F. Good.

1974. Tracers in field measurements of water, milk, and thyroxine metabolism of tropical ruminants, pp. 1-23 in Tracer Techniques in Tropical Animal Production. International Atomic Energy Agency, Vienna.

Macfarlane, W.V., B. Howard, H. Haines, P.J. Kennedy, and C.M. Sharpe.

1971. Hierarchy of water and energy turnover of desert mammals. Nature 234:483-484.

Macfarlane, W.V., B. Howard, G.M.O. Maloiy, and D. Hopcraft.
1972. Tritiated water in field studies of ruminant metabolism in Africa, pp. 83-93 in Isotope Studies on the Physiology of Domestic Animals. International Atomic Energy Agency, Vienna.

Macfarlane, W.V., B. Howard, and R.J.H. Morris.
1966b. Water metabolism of merino sheep shorn during summer. Australian Journal of Agricultural Research 17:219-225.

Macfarlane, W.V., B. Howard, J.F. Morrison, and C.H. Wyndham.
1966c. Content and turnover of water in Bantu miners acclimatized to humid heat. Journal of Applied Physiology 21:978-984.

Macfarlane, W.V., B. Howard, and B.D. Siebert.
1967. Water metabolism of merino and border Leicester sheep grazing saltbush. Australian Journal of Agricultural Research 18:947-958.

1969. Tritiated water in the measurement of milk intake and tissue growth of ruminants in the field. Nature 221:578-579.

Macfarlane, W.V., R.J.H. Morris, and B. Howard.
1963. Turnover and distribution of water in desert camels, sheep, cattle, and kangaroos. Nature 197:270-271.

Macfarlane, W.V., and B.D. Siebert.
1967. Hydration and dehydration of desert camels. Australian Journal of Experimental Biology 45:29.

MacMillen, R.E., and E.A. Christopher.
1975. The water relations of two populations of noncaptive desert rodents, pp. 117-137 in N.F. Hadley, ed., Environmental Physiology of Desert Organisms. Dowden, Hutchison and Ross, Stroudsburg, Pa.

MacMillen, R.E., and D.S. Hinds.
1983. Water regulatory efficiency in heteromyid rodents: A model and its application. Ecology 64:152-164.

Maloiy, G.M.O., ed.
1972. Comparative Physiology of Desert Animals. Symposia of the Zoological Society of London, no. 31. Academic Press, London.

1979. Comparative Physiology of Osmoregulation in Animals, Vols. 1 and 2. Academic Press, New York.

Maloiy, G.M.O, and D. Hopcraft.
1971. Thermoregulation and water relations of two East African antelopes, the hartebeest and impala. Comparative Biochemistry and Physiology 38A:525-534.

Maloiy, G.M.O., W.V. Macfarlane, and A. Shkolnik.
1979. Mammalian herbivores, pp. 185-209 in G.M.O. Maloiy, ed., Comparative Physiology of Osmoregulation in Animals, Vol. 2. Academic Press, New York.

Maltz, E., and A. Shkolnik.
1980. Milk production in the desert: Lactation and water economy in the black bedouin goat. Physiological Zoology 53:12-18.

Marcuzzi, G., and V. Santoro.
1959. Indagini sul ricambio idrico del *Tenebrio molitor* mediante acqua tritiata. Ricerca Scientifica 29:2576-2581.

Martin, J.R., and J.J. Koranda.
1972. Biological half-life studies of tritium in chronically exposed kangaroo rats. Radiation Research 50:426-440.

Mayes, K.R., and D.M. Holdrich.
1975. Water exchange between woodlice and moist environments, with particular reference to *Oniscus asellus*. Comparative Biochemistry and Physiology 51A:295-300.

McClintock, R., and N. Lifson.
1957. Applicability of the $D_2O$-18 method to the measurement of the total carbon dioxide content of obese mice. Journal of Biological Chemistry 226:153-156.

1958a. Determination of the total carbon dioxide output of rats by the $D_2O$-18 method. American Journal of Physiology 192:76-78.

1958b. $CO_2$ output of mice measured by $D_2O$-18 under conditions of isotope re-entry into the body. American Journal of Physiology 195:721-725.

McEwan, E.H., and P.E. Whitehead.
1971. Measurement of the milk intake of reindeer and caribou calves using tritiated water. Canadian Journal of Zoology 49:443-447.

Merker, G.P., and K.A. Nagy.
1984. Energy utilization by free-ranging *Sceloporus virgatus* lizards. Ecology 65:575-581.

*Literature Cited*

Minnich, J.E.
1976. Water procurement and conservation by desert reptiles in their natural environment. Israel Journal of Medical Sciences 12:740-758.

1977. Adaptive responses in the water and electrolyte budgets of native and captive desert tortoises, *Gopherus agassizii*, to chronic drought, pp. 102-129 in Proceedings of the Desert Tortoise Council Symposium.

1979. Reptiles, pp. 391-641 in G.M.O. Maloiy, ed., Comparative Physiology of Osmoregulation in Animals, Vol. 1. Academic Press, New York.

Minnich, J.E., and V.H. Shoemaker.
1970. Diet, behavior, and water turnover in the desert iguana, *Dipsosaurus dorsalis*. American Midland Naturalist 84:496-509.

1972. Water and electrolyte turnover in a field population of the lizard, *Uma scoparia*. Copeia 1972:650-659.

Minnich, J.E., and M.R. Ziegler.
1977. Water turnover of free-living gopher tortoises, *Gopherus polyphemus*, in central Florida, pp. 130-151 in Proceedings of the Desert Tortoise Council Symposium.

Moore, F.D., J.M. Hartsuck, R.M. Zollinger, and J.E. Johnson.
1968. Reference models for clinical studies by isotope dilution. Annals of Surgery 168:671-700.

Morris, K.D., and S.D. Bradshaw.
1981. Water and sodium turnover in coastal and inland populations of the ash-grey mouse, *Pseudomys albocinereus* (Gould), in western Australia. Australian Journal of Zoology 29:519-533.

Morris, R.J.H., B. Howard, and W.V. Macfarlane.
1962. Interaction of nutrition and air temperature with water metabolism of merino wethers shorn in winter. Australian Journal of Agricultural Research 13:330-334.

Morrison, D.A., and K.J. Gallagher.
1984. Possible use of water turnover rates to estimate food intake in lizards. British Journal of Herpetology 6:407-410.

Morton, S.R.
1980. Field and laboratory studies of water metabolism in *Sminthopsis crassicaudata* (Marsupialia:Dasyuridae). Australian Journal of Zoology 28:213-227.

Motais, R., and J. Isaia.
1972. Temperature dependence of permeability to water and to sodium of the gill epithelium of the eel, *Anguilla anguilla*. Journal of Experimental Biology 56:587-600.

Motais, R., J. Isaia, J.C. Rankin, and J. Maetz.
1969. Adaptive changes of the water permeability of the teleostean gill epithelium in relation to external salinity. Journal of Experimental Biology 51:529-546.

Mullen, R.K.
1970. Respiratory metabolism and body water turnover rates of *Perognathus formosus* in its natural environment. Comparative Biochemistry and Physiology 32:259-265.

1971. Energy metabolism and body water turnover rates of two species of free-living kangaroo rats, *Dipodomys merriami* and *Dipodomys microps*. Comparative Biochemistry and Physiology 39A:379-390.

Nagy, K.A.
1972. Water and electrolyte budgets of a free-living desert lizard, *Sauromalus obesus*. Journal of Comparative Physiology 79:39-62.

1980. $CO_2$ production in animals: Analysis of potential errors in the doubly labeled water method. American Journal of Physiology 238:R466-R473.

1982. Field studies of water relations, pp. 484-501 in C. Gans and F.H. Pough, eds., Biology of the Reptilia, Vol. 12. Academic Press, New York.

1983a. Ecological energetics, pp. 24-54 in R.B. Huey, E.R. Pianka, and T.W. Schoener, eds., Lizard Ecology. Harvard University Press, Cambridge, Mass.

1983b. The doubly labeled water ($^3HH^{18}O$) method: A guide to its use. University of California, Los Angeles, Pub. no. 12-1417.

1987. Field metabolic rate and food requirement scaling in mammals and birds. Ecological Monographs 57:111-128.

Nagy, K.A., and D.P. Costa.
1980. Water flux in animals: Analysis of potential errors in the tritiated water method. American Journal of Physiology 238:R454-R465.

Nagy, K.A., R.B. Huey, and A.F. Bennett.
1984a. Field energetics and foraging mode of Kalahari lacertid lizards. Ecology 65:588-596.

Nagy, K.A., and R.W. Martin.
1985. Field metabolic rate, water flux, food consumption, and time budget of koalas, *Phascolarctos cinereus* (Marsupialia: Phascolarctidae) in Victoria. Australian Journal of Zoology 33:655-665.

Nagy, K.A., and P.A. Medica.
1986. Physiological ecology of desert tortoises in southern Nevada. Herpetologica 42:73-92.

Nagy, K.A., and K. Milton.
1979a. Energy metabolism and food consumption by wild howler monkeys (*Alouatta palliata*). Ecology 60:475-480.

1979b. Aspects of dietary quality, nutrient assimilation, and water balance in wild howler monkeys (*Alouatta palliata*). Oecologia 39:249-258.

Nagy, K.A., and G.G. Montgomery.
1980. Field metabolic rate, water flux, and food consumption in three-toed sloths (*Bradypus variegatus*). Journal of Mammalogy 61:465-472.

Nagy, K.A., R.S. Seymour, A.K. Lee, and R. Braithwaite.
1978. Energy and water budgets in free-living *Antechinus stuartii* (Marsupialia:Dasyuridae). Journal of Mammalogy 59:60-68.

Nagy, K.A., and V.H. Shoemaker.
1975. Energy and nitrogen budgets of the free-living desert lizard *Sauromalus obesus*. Physiological Zoology 48:252-262.

1984. Field energetics and food consumption of the Galapagos marine iguana, *Amblyrhnchus cristatus*. Physiological Zoology 57:281-290.

Nagy, K.A., V.H. Shoemaker, and W.R. Costa.
1976. Water, electrolyte, and nitrogen budgets of jackrabbits (*Lepus californicus*) in the Mojave Desert. Physiological Zoology 49:351-363.

Nagy, K.A., W.R. Siegfried, and R.P. Wilson.
1984b. Energy utilization by free-ranging jackass penguins, *Spheniscus demersus*. Ecology 65:1648-1655.

Nagy, K.A., and G.C. Suckling.
1985. Field energetics and water balance of sugar gliders, *Petaurus breviceps* (Marsupialia:Petauridae). Australian Journal of Zoology 33:683-691.

Nicol, S.C.
1978. Rates of water turnover in marsupials and eutherians: A comparative review, with new data on the Tasmanian devil. Australian Journal of Zoology 26:465-473.

Nicolson, S.W., and J.P. Leader.
1974. The permeability to water of the cuticle of the larva of *Opifex fuscus* (Hutton) (Diptera, Culicidae). Journal of Experimental Biology 60:593-603.

Nobel, P.S.
1980. Water vapor conductance and $CO_2$ uptake for leaves of a $C_4$ desert grass, *Hilaria rigida*. Ecology 61:252-258.

1983. Biophysical plant physiology and ecology. W.H. Freeman, San Francisco.

Obst, B.S., K.A. Nagy, and R.E. Ricklefs.
1987. Energy utilization by Wilson's storm-petrel (*Oceanites oceanicus*). Physiological Zoology 60:200-210.

Oddershede, I., and R.S. Elizondo.
1980. Water content, water turnover, and water half-life during cold acclimation in the rhesus monkey, *Macaca mulatta*. Canadian Journal of Physiology and Pharmacology 58:34-39.

Ohata, C.A., L.K. Miller, and D.F. Holleman.
1975. Water metabolism and body composition in the northern fur seal, p. 15 in University of California, Santa Cruz, Conference on the Biology and Conservation of Marine Mammals.

Ohmart, R.D., T.E. Chapman, and L.Z. McFarland.
1970. Water turnover in roadrunners under different environmental conditions. The Auk 87:787-793.

Olsson, K.E.
1970. Determination of total body water and its turnover rate: A methodological study with tritium-labeled water. Acta Chirurgica Scandinavica 136:647.

Ortiz, C.L., D.P. Costa, and B.J. Le Boeuf.
1978.  Water and energy flux in elephant seal pups fasting under natural conditions. Physiological Zoology 52:166-178.

Ortiz, C.L., B.J. LeBoeuf, and D.P. Costa.
1984.  Milk intake of elephant seal pups: An index of parental investment. American Naturalist 124:416-422.

Paganelli, C.V., and A.K. Solomon.
1957.  The rate of exchange of tritiated water across the human red-cell membrane. Journal of General Physiology 41:259-277.

Parry, G.
1955.  Urine production by the antennal glands of *Palaemonetes varians*. Journal of Experimental Biology 32:408-422.

Payan, P., L. Goldstein, and R.P. Forster.
1973.  Gills and kidneys in ureosmotic regulation in euryhaline skates. American Journal of Physiology 224:367-372.

Payan, P., and J. Maetz.
1971.  Balance hydrique chez les élasmobranches: arguments en faveur d'un controle endocrinien. General and Comparative Endocrinology 16:535-554.

Peitz, B.
1974.  Influence of the estrus cycle and estrogen on total body water and biological half-life of water in the rat, spiny mouse, and prairie dog. American Zoologist 14:1244.

Peters, R.H.
1983.  The Ecological Implications of Body Size. Cambridge University Press, Cambridge, England.

Phillips, R.W., and K.L. Knox.
1969.  Water kinetics in enteric disease of neonatal calves. Journal of Dairy Science 52:1664-1668.

Phillips, R.W., K.L. Knox, W.A. House, and H.N. Jordan.
1969.  Metabolic response in sheep chronically exposed to 6,200 m simulated altitude. Federation Proceedings 28:974.

Pinet, J.M., L.A. Buscarlet, P. Garrigues, and F. Reitz.
1982.  Renouvellement de l'eau corporelle et bilan energetique chez la perdix grise, *Perdix perdix*. Acta Oecologica/Oecologia Applicata 3:79-94.

Pinshow, B., A.A. Degen, and P.U. Alkon.
1983.   Water intake, existence energy, and response to water deprivation in the sand partridge, *Ammoperdix heyi*, and the chukar, *Alectoris chukar*: two phasianids of the Negev Desert. Physiological Zoology 56:281-289.

Pinson, E.A., and W.H. Langham.
1957.   Physiology and toxicology of tritium in man. Journal of Applied Physiology 10:108-126.

Potter, G.D., G.M. Vattuone, and D.R. McIntyre.
1972.   Metabolism of tritiated water in the dairy cow. Health Physics 22:405-409.

Potts, W.T.W., and F.B. Eddy.
1973.   The permeability to water of the eggs of certain marine teleosts. Journal of Comparative Physiology 82:305-315.

Potts, W.T.W., and W.R. Fleming.
1970.   Effects of prolactin and divalent ions on permeability to water of *Fundulus kansae*. Journal of Experimental Biology 53:317-327.

Potts, W.T.W., M.A. Foster, P.P. Rudy, and G.P. Howells.
1967.   Sodium and water balance in the cichlid teleost, *Tilapia mossambica*. Journal of Experimental Biology 47:461-470.

Potts, W.T.W., M.A. Foster, and J.W. Stather.
1970.   Salt and water balance in salmon smolts. Journal of Experimental Biology 52:553-564.

Potts, W.T.W., and P.P. Rudy.
1969.   Water balance in eggs of Atlantic salmon, *Salmo salar*. Journal of Experimental Biology 50:223-237.

1972.   Aspects of osmotic and ionic regulation in the sturgeon. Journal of Experimental Biology 56:703-715.

Prosser, C.L.
1973.   Comparative Animal Physiology, 3rd ed. Cambridge University Press, Cambridge, England.

Purdue, J.R., and H. Haines.
1977.   Salt-water tolerance and water turnover in the snowy plover. The Auk 94:248-255.

*Literature Cited*

Ranjhan, S.K., A.P. Kalanidhi, T.K. Gosh, U.B. Singh, and K.K. Saxena.
1982. Body composition and water metabolism in tropical ruminants, using tritiated water, pp. 117-132 in Use of Tritiated Water in Studies of Production and Adaptation in Ruminants. International Atomic Energy Agency, Vienna.

Rath, E.A., and S.W. Thenen.
1979. Use of tritiated water for measurement of 24-hour milk intake in suckling lean and genetically obese mice. Journal of Nutrition 109:840-847.

Ray, C.T., and G.E. Burch.
1959. Relationship of equilibrium of distribution, biologic decay rates, and space and mass of H-3, Cl-36, and Rb-86 observed simultaneously in a comfortable and in a hot and humid environment, in control subject and in patient with congestive heart failure. Journal of Laboratory and Clinical Medicine 53:69-88.

Reese, J.B., and H. Haines.
1978. Effects of dehydration on metabolic rate and fluid distribution in the jackrabbit, *Lepus californicus*. Physiological Zoology 51:155-165.

Richards, G.C.
1979. Variation in water turnover by wild rabbits, *Oryctolagus cuniculus*, in an arid environment due to season, age group, and reproductive condition. Australian Wildlife Research 6:289-296.

Richmond, C.R., W.H. Langham, and T.T. Trujillo.
1962. Comparative metabolism of tritiated water by mammals. Journal of Cellular and Comparative Physiology 59:45-53.

Richmond, C.R., T.T. Trujillo, and D.W. Martin.
1960. Volume and turnover of body water in *Dipodomys deserti* with tritiated water. Proceedings of the Society of Experimental Biology 104:9-11.

Ricklefs, R.E., D.D. Roby, and J.B. Williams.
1986. Daily energy expenditure by adult Leach's Storm-Petrels during the nesting cycle. Physiological Zoology 59:649-660.

Ricklefs, R.E., and J.B. Williams.
1984. Daily energy expenditure and water turnover rate of adult European starlings (*Sturnus vulgaris*) during the nesting cycle. The Auk 101:707-716.

Roberts, J.E., K.D. Fisher, and T.H. Allen.
1958. Tracer method for estimating total water exchange in man. Physics in Medicine and Biology 3:7-15.

Robinson, G.D.
1982. Water fluxes and urine production in blue crabs (*Callinectes sapidus*) as a function of environmental salinity. Comparative Biochemistry and Physiology 71A:407-412.

Robinson, G.D., and W.A. Dunson.
1976. Water and sodium balance in the estuarine diamondback terrapin (*Malaclemys*). Journal of Comparative Physiology 105:129-152.

Robinson, G.D., and J.E. Ewig.
1983. Water fluxes in red efts, prolactin-injected efts, and newts (*Notophthalmus viridescens*) exposed to freshwater. Comparative Biochemistry and Physiology 74A:927-931.

Roesijadi, G.
1978. Water turnover rates in the megalopa and crab stages I-V of *Pinnixa occidentalis*. Comparative Biochemistry and Physiology 59A:259-260.

Romero, J.J., R. Casas, and R.L. Baldwin.
1975. A technique for estimating milk production in rats. Journal of Nutrition 105:413-420.

Rooke, I.J., S.D. Bradshaw, and R.A. Langworthy.
1983. Aspects of the water, electrolyte, and carbohydrate physiology of the silvereye, *Zosterops lateralis* (Aves). Australian Journal of Zoology 31:695-704.

Rouffignac, C. de, and F. Morel.
1966. Étude comparée du renouvellement de l'eau chez quatre espèces de rongeurs, dont deux espèces d'habitat désertique. Journal de Physiologie, Paris 58:309.

Rubsamen, K., and W.V. Engelhardt.
1975. Water metabolism in the llama. Comparative Biochemistry and Physiology 52A:595-598.

Rubsamen, K. and W.V. Engelhardt
1982. Water economy and temperature regulation in the rock hyrax, pp. 207-216 in Use of Tritiated Water in Studies of Production and Adaptations in Ruminants. International Atomic Energy Agency, Vienna.

Rubsamen, K., R. Keller, H. Lawrenz, and W.V. Engelhardt.
1979. Water and energy metabolism in the rock hyrax (*Procavia habessinica*). Journal of Comparative Physiology 131:303-309.

Rudy, P.P.
1967. Water permeability in selected decapod crustacea. Comparative Biochemistry and Physiology 22:581-589.

Rudy, P.P., and R.C. Wagner.
1970. Water permeability in the Pacific hagfish, *Polistotrema stouti*, and the staghorn sculpin, *Leptocottus armatus*. Comparative Biochemistry and Physiology 34:399-403.

Russel, A.J.F., J.Z. Foot, and D.M. McFarlane.
1982. Use of tritiated water for estimating body composition in grazing ewes, pp. 45-56 in Use of Tritiated Water in Studies of Production and Adaptation in Ruminants. International Atomic Energy Agency, Vienna.

Schierwater, B., and H. Klingel.
1985. Food digestibility and water requirements in the Djungarian hamster, *Phodopus sungorus*. Zeitschrift für Saugetierkunde 50:35-39.

Schloerb, P.R., B.J. Friis-Hansen, I.S. Edelman, A.K. Solomon, and F.D. Moore.
1950. The measurement of total body water in the human subject by deuterium oxide dilution, with a consideration of the dynamics of deuterium distribution. Journal of Clinical Investigation 29:1296-1310.

Schmidt-Nielsen, K.
1964. Desert Animals: Physiological Problems of Heat and Water. Oxford University Press, New York.

1983. Animal Physiology: Adaptation and Environment, 3rd ed. Cambridge University Press, Cambridge, England.

1984. Scaling: Why is Animal Size so Important? Cambridge University Press, Cambridge, England.

Schoeller, D.A., and E. van Santen.
1982. Measurement of energy expenditure in humans by doubly labeled water method. Journal of Applied Physiology 53:955-959.

Schoeller, D.A., and P. Webb.
1984. Five-day comparison of the doubly labeled water method with respiratory gas exchange. American Journal of Clinical Nutrition 40:153-158.

Schultheiss, H., W. Hanke, and J. Maetz.
1972. Hormonal regulation of the skin: Diffusional permeability to water during development and metamorphosis of *Xenopus laevis* Daudin. General and Comparative Endocrinology 18:400-404.

Seefeldt, S.L., and T.E. Chapman.
1979. Body water content and turnover in cats fed dry and canned rations. American Journal of Veterinary Research 40:183-185.

Shaw, J.
1955. The permeability and structure of the cuticle of the aquatic larva of *Sialis lutaria*. Journal of Experimental Biology 32:330-352.

Shoemaker, V.H., and K.A. Nagy.
1977. Osmoregulation in amphibians and reptiles. Annual Review of Physiology 39:449-471.

1984. Osmoregulation in the Galapagos marine iguana, *Amblyrhynchus cristatus*. Physiological Zoology 57:291-300.

Shoemaker, V.H., K.A. Nagy, and W.R. Costa.
1976. Energy utilization and temperature regulation by jackrabbits (*Lepus californicus*) in the Mojave Desert. Physiological Zoology 49:364-375.

Sicard, B., M. Navajas y Navarro, T. Jacquart, F. Lachiver, and H. Croset.
1985. Métabolisme hydrique de populations de *Mus musculus domesticus* et de *Mus spretus* Lataste soumises à divers régimes hydriques. Comptes Rendus des Seances de l'Academie des Sciences, Paris 300:699-704.

Siebert, B.D.
1971. Growth and water metabolism of cows and progeny on fertilized and unfertilized tropical pastures. Australian Journal of Agricultural Research 22:415-428.

Siebert, B.D., and W.V. Macfarlane.
1969. Body water content and water turnover of tropical *Bos indicus*, *Bibos banteng*, and *Bubalus bubalis*. Australian Journal of Agricultural Research 20:613-622.

1971. Water turnover and renal function in dromedaries in the desert. Physiological Zoology 44:225-240.

Sigal, M.D., and L.G. Arlian.
1982. Water balance of the social insect *Formica exsectoides* (Hymenoptera: Formicidae) and its ecological implications. Physiological Zoology 55:355-366.

Skadhauge, E., and S.D. Bradshaw.
1974. Saline drinking and cloacal excretion of salt and water in the zebra finch. American Journal of Physiology 227:1263-1267.

Smith, A.P., K.A. Nagy, M.R. Fleming, and B. Green.
1982. Energy requirements and water turnover in free-living Leadbeater's possums, *Gymnobelideus leadbeateri* (Marsupialia:Petauridae). Australian Journal of Zoology 30:737-749.

Smith, P.G.
1969. Ionic relations of *Artemia salina* (L.). II. Fluxes of sodium, chloride, and water. Journal of Experimental Biology 51:739-757.

Smith, R.I.
1964. $D_2O$ uptake rate in two brackish-water nereid polychaetes. Biological Bulletin 126:142-149.

1967. Osmotic regulation and adaptive reduction of water permeability in a brackish water crab, *Rhithropanopeus harrisi* (Brachyura, Xanthidae). Biological Bulletin 133:643-658.

1970a. The apparent water permeability of *Carcinus maenas* (Crustacea Brachyura, Portunidae) as a function of salinity. Biological Bulletin 139:351-362.

1970b. Chloride regulation at low salinities by *Nereis diversicolor* (Annelida, Polychaeta). II. Water fluxes and apparent permeability to water. Journal of Experimental Biology 53:93-100.

Smith, R.I., and P.P. Rudy.
1972. Water exchange in the crab *Hemigrapsus nudus* measured by use of deuterium and tritium oxides as tracers. Biological Bulletin 143:234-246.

Smits, A.W.
1985. Correlates of activity, diet, and body water flux in the chuckwalla lizard *Sauromalus hispidus*. Physiological Zoology 58:166-174.

Snyder, W.S., B.R. Fish, S.R. Bernard, M.R. Ford, and J.R. Muir.
1968. Urinary excretion of tritium, following exposure of man to HTO—a two-exponential model. Physics in Medicine and Biology 13:547-559.

Solomon, A.K.
1960. Pores in the cell membrane. Scientific American 203:146-156.

Soman, S.D., and T.M. Krishnamoorthy.
1973. Body water turnover rates in *Anadara granosa*. Current Science 42:453-456.

Springell, P.H.
1968. Water content and water turnover in beef cattle. Australian Journal of Agricultural Research 19:129.

Staddon, B.W.
1966. Permeability of water of cuticles of some adult water bugs. Journal of Experimental Biology 44:69-76.

Stephenson, A.H., and J.E. Minnich.
1974. Water turnover in free-living song sparrows in southeastern Wisconsin. American Zoologist 14:1257.

Streit, B.
1980. Untersuchungen zum Wasseraustausch (mittels $^3H_2O$) zwischen Süsswassertieren und ihrer Umgebung. Revue Suisse de Zoologie 87:927-936.

1982. Water turnover rates and half-life times in animals studied by use of labelled and non-labelled water. Comparative Biochemistry and Physiology 72A:445-454.

Subramanian, A.
1975. Sodium and water permeabilities in selected crustacea. Physiological Zoology 48:398-403.

Takeda, H.
1982. Comparative metabolism of tritium in rat after single ingestion of some tritiated-labeled organic compounds vs. tritium-labeled water. Journal of Radiation Research 23:345-357.

Taplin, L.E.
1985. Sodium and water budgets of the fasted estuarine crocodile, *Crocodylus porosus*, in sea water. Journal of Comparative Physiology B155:501-514.

Thomas, D.H., and J.G. Phillips.
1975. Studies in avian adrenal steroid function. II. Chronic adenalectomy and the turnover of tritiated water in domestic ducks, *Anas platyrhynchos*. General and Comparative Endocrinology 26:404-411.

Thompson, R.C.
1952. Studies of metabolic turnover with tritium as a tracer. I. Gross studies on the mouse. Journal of Biological Chemistry 197:81-87.

1953. Studies on metabolic turnover with tritium as a tracer. II. Gross studies on the rat. Journal of Biological Chemistry 200:731-743.

Tucker, J.S., and F.L. Harrison.
1974. The incorporation of tritium in the body water and organic matter of selected marine invertebrates. Comparative Biochemistry and Physiology 49A:387-397.

Ueno, Y., and K. Kawamura.
1975. Study on the elimination of tritium in mice injected with tritiated water. Health Physics 29:413-415.

Utter, J.M.
1971. Daily energy expenditures of free-living purple martins (*Progne subis*) and mockingbirds (*Mimus polyglottos*), with a comparison of two northern populations of mockingbirds. Ph.D. dissertation, Rutgers University, New Brunswick, N.J.

Utter, J.M., and E.A. LeFebvre.
1973. Daily energy expenditure of purple martins (*Progne subis*) during the breeding season: Estimates using $D_2O$-18 and time-budget methods. Ecology 54:597-604.

Van Hook, R.I., and S.I. Deal.
1973. Tritium uptake and elimination by tissue-bound and body water components in crickets (*Acheta domesticus*). Journal of Insect Physiology 19:681-687.

Walter, A., and M.R. Hughes.
1978. Total body water volume and turnover rate in fresh-water and sea-water adapted glaucous-winged gulls, *Larus glaucescens*. Comparative Biochemistry and Physiology 61A:233-237.

Weathers, W.W., and K.A. Nagy.
1980. Simultaneous doubly labeled water (tritium and oxygen-18 labeled) and time-budget estimates of daily energy expenditure in *Phainopepla nitens*. The Auk 97:861-867.

1984. Daily energy expenditure and water flux in black-rumped waxbills (*Estrilda troglodytes*). Comparative Biochemistry and Physiology 77A:453-458.

Wesley, D.E., K.L. Knox, and J.G. Nagy.
1970. Energy flux and water kinetics in young pronghorn antelope. Journal of Wildlife Management 34:908-912.

Wharton, G.W., and T.L. Devine.
1968. Exchange of water between a mite, *Laelaps echidnina*, and the surrounding air under equilibrium conditions. Journal of Insect Physiology 14:1303-1318.

Wheeler, J.K., A.A. Moghissi, B.F. Rehnberg, and M.C. Colvin.
1972. Comparison between the biological half-life of tritiated luminous compound with that of tritiated water in rats and cats. Health Physics 22:35-38.

Whittow, G.C.
1986. Energy metabolism, pp. 253-268 in P.D. Sturkie, ed., Avian Physiology, 4th ed. Springer-Verlag, New York.

Williams, C.K., and B. Green.
1982. Ingestion rates and aspects of water, sodium and energy metabolism in caged swamp buffalo, *Bubalus bubalis*, from isotope dilution and materials balances. Australian Journal of Zoology 30:779-790.

Williams, C.K., and M.G. Ridpath.
1982. Rates of herbage ingestion and turnover of water and sodium in feral swamp buffalo, *Bubalus bubalis*, in relation to primary production in a cyperaceous swamp in monsoonal northern Australia. Australian Wildlife Research 9:397-408.

Williams, J.B.
1985. Validation of the doubly labeled water technique for measuring energy metabolism in starlings and sparrows. Comparative Biochemistry and Physiology 80A:349-353.

Williams, J.B., and K.A. Nagy.
1984a. Validation of the doubly labeled water technique for measuring energy metabolism in savannah sparrows. Physiological Zoology 57:325-328.

1984b. Daily energy expenditure of savannah sparrows: Comparison of time-energy budget and doubly labeled water estimates. The Auk 101:221-229.

1985a. Daily energy expenditure by female savannah sparrows feeding nestlings. The Auk 102:187-190.

1985b. Water flux and energetics of nestling savannah sparrows in the field. Physiological Zoology 58:515-525.

Wilson, A.D.
1974. Water consumption and water turnover of sheep grazing semi-arid pasture communities in New South Wales. Australian Journal of Agricultural Research 25:339-347.

Withers, P.C.
1983. Energy, water, and solute balance of the ostrich, *Struthio camelus*. Physiological Zoology 56:568-579.

Withers, P.C., G.N. Louw, and J. Henschel.
1980. Energetics and water relations of Namib Desert, southwest Africa, rodents. South African Journal of Zoology 15:131-137.

Wright, D.E.
1982a. Applications of labelled water in animal nutrition and physiology. I. Measurement of individual intakes of grazing animals, pp. 69-76; II. Measurement of milk intake, pp. 77-88 in Use of Tritiated Water in Studies of Production and Adaptation in Ruminants. International Atomic Energy Agency, Vienna.

1982b. Use of labelled water in studies on the nutrition and physiology of grazing ruminants in New Zealand, pp. 91-102 in Use of Tritiated Water in Studies of Production and Adaptations in Ruminants. International Atomic Energy Agency, Vienna.

Wylie, K.F., W.A. Bigler, and G.R. Grove.
1963.   Biological half-life of tritium. Health Physics 9:911-914.

Yates, N.G., W.V. Macfarlane, and R. Ellis.
1971.   Estimation of milk intake and growth of beef calves in the field by using tritiated water. Australian Journal of Agricultural Research 22:291-306.

Yokota, S.D.
1979.   Water, energy and nitrogen metabolism in the desert scorpion, *Paruroctonus mesaensis*. Ph.D. dissertation. University of California, Riverside.

Yousef, M.K.
1971.   Tritiated water turnover rate in desert mammals, pp. 333-341 A.A. Moghissi and M.W. Carter, eds., Tritium. Messenger Graphics, Phoenix, Ariz.

Yousef, M.K., H.D. Johnson, W.G. Bradley, and S.M. Seif.
1974.   Tritiated-water turnover rate in rodents, desert and mountain. Physiological Zoology 47:153-162.

Zervanos, S.M., and G.I. Day.
1977.   Water and energy requirements of captive and free-living collared peccaries. Journal of Wildlife Management 41:527-535.